Operations Research
New Paradigms and Emerging Applications

Editors

Vilda Purutçuoğlu
Professor, Department of Statistics
Middle East Technical University, Ankara, Turkey

Gerhard-Wilhelm Weber
Chairman, Marketing and Economical Engineering
Poznan University of Technology, Poznan, Poland

Hajar Farnoudkia
Department of Business Administration
Başkent University, Ankara, Turkey

CRC Press
Taylor & Francis Group
Boca Raton London New York

CRC Press is an imprint of the
Taylor & Francis Group, an **informa** business
A SCIENCE PUBLISHERS BOOK

First edition published 2023
by CRC Press
6000 Broken Sound Parkway NW, Suite 300, Boca Raton, FL 33487-2742

and by CRC Press
4 Park Square, Milton Park, Abingdon, Oxon, OX14 4RN

Library of Congress Cataloging-in-Publication Data (applied for)

ISBN: 978-1-032-34926-8 (hbk)
ISBN: 978-1-032-34929-9 (pbk)
ISBN: 978-1-003-32450-8 (ebk)

DOI: 10.1201/9781003324508

Typeset in Times New Roman
by Radiant Productions

Foreword

This book aims to provide Operation Research (OR) applications under twelve chapters organized in the form of four parts according to the main topics considered in these chapters. These four parts are optimization; data mining and clustering; business, science and finance; and medical application. Hereby, the continuous numbering of chapters is applied in those listed parts.

Accordingly, **Part 1** is allocated to Operation Research in optimization, given in three chapters.

Chapter 1 proposes a parameter estimation way for partially nonlinear problems which are semi-parametric regression models and the extension of the partially linear model that has gained importance in the statistical literature. These models are employed when the non-parametric regression model does not perform well. In this regard, firstly, nonlinear least square estimation is established based on the Taylor expansion of the nonlinear function. Then, a kernel-based bridge problem is employed to estimate the non-parametric component of the model. In the end, the optimization method is applied to choose the best estimation.

Chapter 2 provides a glimpse on the contributions and challenges towards more environmentally-friendly road traffic, by reviewing academic studies on how Operation Research has been used in controlling the complex transportation network since connected and analyzing traffic-related impacts, especially regarding environmental and air quality, of automated vehicles are not fully deployed yet on the roads. Operation Research in this matter is used to plan for future challenges and major impacts can be expected, as well.

Chapter 3 of this part represents some case studies for selecting suppliers and portfolio investment schemes by addressing the application of the multi-criteria method on discrete variables that are very important in complex decision-making problems.

Part II is devoted to new applications of Operations Research to Data Mining and Clustering and consists of three chapters. *Chapter 4* in this part is based on a survey about the dimension reduction methods such as clustering and principal component analysis (PCA). The main aim of this chapter is to investigate a semi-definite programming model that provides an effective solution to problems related to both PCA and clustering methods.

Chapter 5 of this part discusses some practical techniques for different types of clustering by using the formulation of the problems as an optimization method. Here, the clustering problems can be represented in terms of a real function of several real variables, and a set of arguments that give an optimal clustering.

Chapter 6, which is the last chapter of data mining and clustering portion, is dedicated to the application of these methods to meteorological data by presenting a review of the analysis. Then, as the illustration of the proposal approaches, the meteorological data are chosen. In the analyses, the data preparation and pre-processing are also explained in detail besides the clustering and the modeling in terms of the findings.

Part III of this book is based on operation research in business, science, and Finance in three distinct chapters. *Chapter 7* is about the foundations of market-making via stochastic optimal control. Market-making is a type of high-frequency trading that implies the quantitative trading of a short portfolio holding period. This chapter covers several results obtained by traditional techniques of stochastic optimal control related to market-making which is a significant component of financial research. *Chapter 8* is about the decision aid to drive the network with a better management of the system while there could be a conflict in used criteria such as the minimization of the cost and the maximization of the security simultaneously. Furthermore, there can be other objectives that may appear unexpectedly. So, this chapter is dedicated to explain why the formulation of the multi-objective network flow problem is a necessity and how it can be done.

Chapter 9 of the third part of this book is about operation research application in decision-making in finance focusing on the behavior of financial problems which are based on the investors' behavior introduced on sentiment. It also compares the forecasting performances of sentiments index by using different mathematical models.

Part IV represents operation research in medical application in three chapters. *Chapter 10* starts with a study covering necessary information about an algorithm and a stability approach for the acute inflammatory response dynamic model. Generally, the filamentary response wipes out the pathogens from the body and repairs the healthy case. Recently, mathematical models are being used to provide essential insights into the dynamics of the inflammatory response. On the other side, nonlinear dynamics have gained high importance in many areas that can describe the complicated conceptions within details. This study provides a numerical approximation to the complicated systems via nonlinear differential equations.

Chapter 11 titled Bayesian inference for the undirected network tries to estimate the conditional dependence between genes by using a Monte Carlo algorithm in case the number of parameters exceeds the number of observations. Here, the parameter in the precision matrix does not have a fixed dimension in each iteration. To overcome the problem, a reversible jump Markov chain Monte Carlo method, which is one of the OR algorithms optimizing both the fitted model to the data and the estimated parameters, is proposed. Furthermore, its alternatives such as Gibbs sampling, and Carlin Chibs methods, which are other advanced inference methods based on optimization of some score functions, are introduced. Moreover, as one of the powerful tools for modeling the biological network, the copula is introduced with its special type named the vine copula which tries to simplify the multivariate complicated model into the bivariate model. In the light of the vine copula, each undirected connection can be represented independently with the best copula family.

Finally, *Chapter 12*, the last chapter of the book, presents the comparison and transfer of the EMG data from two stations. In the compression of these signal data, two optimization techniques, namely, dynamic cosine transformation and principal component analysis, are implemented. Then, their performances are compared via distinct accuracy measures once their compressed values are classified by different clustering approaches that are well-known in the field of operation research.

As a result, we consider that this book can present the variety of applications of operation research techniques in different problems. We hope that the references used in each part can be also useful for the new researchers while deeply learning the theoretical aspects of the selected methods. Hence, we hope that the book can open new avenues for novel researches.

Preface

This original book on Operational Research: New Paradigms and Emerging Applications surveys and details newest technology in analytics and intelligence computing, artificial intelligence (AI) and operational research (OR) which are reducing the dimensions of data coverage as well as variables worldwide. This compendium discusses code of intelligent optimization which can be applied in various branches of optimization, data mining and clustering, economic/finance and medical applications. Involving modern and emerging techniques of OR and AI, and applying them together with strong and evolutionary algorithms to real-life problems for strategical, but also daily applications, this compendium elaborates all areas of OR results, methods and applications. By the rich diversity of this handbook, the state-of-the-art developments in quickly advancing key technologies are covered. In this way, with our reference work we hope to be useful for students and emerging scientists of engineering and science, management and economics, social science and the art, for researchers and scholars who are employed in OR supported industries, for decision-makers and designers of tomorrow's World.

We editors hope that the chosen subjects and picked areas reflect a core sample of international OR research facing emerging, challenging, complex and even long-enduring problems of our environment and their field in economics and finance, natural sciences and engineering, healthcare and medicine, industry and city planning, through the results and tools of OR-Analytics. We are very grateful to the publishing house of CRC Press for the honor of accommodating this front running project in intelligence and science, operations and implementation. We convey particular thanks to the directors, editors and managers of CRC Press as well as to its editorial management and team, for their steady interest, care and encouragement, recommendations, support and guide in every respect. We thank our respected authors by their diligent work and readiness to share their newest findings, insights and results with over international community. Now we hope that the research of our authors collected and edited by us will be an inspiration for cooperation and joint implementation, improvement and friendship at the global stage and premium level.

Acknowledgement

I would like to thank deeply to Prof. Dr. Gerhard-Wilhelm Weber for his great support throughout my academic career, and Dr. Hajar Farnoudkia for her hardworking and willingness in this book. Finally, my utmost sincerest thanks go to my parents, my mother Yurdagül Purutçuoğlu, my father Ahmet Purutçuoğlu, my sister Prof. Dr. Eda Purutçuoğlu and my sweat daughter Vera to their big love and support.

—Vilda Purutçuoğlu

I would like to cordially thank Prof. Dr. Vilda Purutçuoğlu for her vision, care and inspiration, and Dr. Hajar Farnoudkia for her dedicated hard-work. All of these made this book become so beautiful and such a success.

—Gerhard-Wilhelm Weber

I want to thank my PhD advisor, Prof. Dr. Vilda Purutçuoğlu for her passion and knowledge who taught me how to study, search and write. I owe a big part of my success to her. Also I think here is a good opportunity to thank Assoc. Prof. Dr. Babek Erdebill who encouraged and helped me to be one of the academic world members.

—Hajar Farnoudkia

All editors also appreciate the great contribution of all authors as they accepted to share their knowledge and researches with them. Finally, they like to convey their sincerest thanks to Directors, Editors, Managers of CRC Press and Vijay Primlani, the acquisitions editor of CRC Press, and his team for their supports and guide in every respect.

Contents

Part III: Operation Research in Business Science and Finance

Part IV: Operation Research in Medical Application

Contributors

Akan, Erhan
Turkish Aerospace Industries
Ankara, Turkey

Aktaş Dinçer, Hayriye
Fatih Sultan Mehmet Vakıf University
İstanbul, Turkey

Arı, Fikret
Ankara University
Ankara, Turkey

Bandeira, Jorge
University of Aveiro
Aveiro, Portugal

Batmaz, İnci
Middle East Technical University
Ankara, Turkey

Belkacem, Salima Nait
University of M'hamed Bougara
Boumerdès, Algeria

Can Erkuş, Ekin
Middle East Technical University
Ankara, Turkey

Farnoudkia, Hajar
Başkent University
Ankara, Turkey

Farzin Asanjan, Mahdieh
Middle East Technical University
Ankara, Turkey

Freitas, Adelaide
University of Aveiro
Aveiro, Portugal

Gökçay, Didem
21yy Ltd.
Ankara, Turkey

Gökçe, Aytül
Ordu University,
Ordu, Turkey

Gomez-Rueda, Mario Sergio
Universidad Pontificia Bolivariana
Medellín, Colombia

Gürbüz, Burcu
Johannes Gutenberg-University
Mainz, Mainz, Germany

İleri, Fatih
Turkish Aerospace Industries
Ankara, Turkey

Kalaycı, Betül
Middle East Technical University
Ankara, Turkey

Lotero, Laura
Universidad Pontificia Bolivariana
Medellín, Colombia

Macedo, Eloisa
University of Aveiro
Aveiro, Portugal

Purutçuoğlu, Vilda
Middle East Technical University
Ankara, Turkey

Savku, Emel
University of Oslo
Oslo, Norway

Somuncuoğlu, Abdullah Nuri
Middle East Technical University
Ankara, Turkey

Taylan, Pakize
Dicle University
Diyarbakır, Turkey

Tchemisova, Tatiana
University of Aveiro
Aveiro, Portugal

Wilhelm Weber, Gerhard
Poznan University of Technology
Poznan, Poland

Yerlikaya Özkurt, Fatma
Atılım University
Ankara, Turkey

I

OPERATION RESEARCH IN OPTIMIZATION

Chapter 1

Kernel Based C-Bridge Estimator for Partially Nonlinear Model

Pakize Taylan

1.1 Introduction

Semi-parametric regression models (SPRMs) [28] deal with regression models which consider the effects of both the parametric and nonparametric regression models simultaneously. They are very helpful for data analysis since they keep the flexibility of nonparametric models and the properties easy interpretation of parametric models to comment baseline function f. Therefore, SPRMs have attracted considerable attention in recent years and have been studied by many researchers interested in data analysis. For this chapter, we consider partially nonlinear models (PNLMs) that are SPRMs and extensions of partially linear models (PLMs) [15] which have been popular in the statistical literature. PNLMs are employed when nonparameric regression does not perform well.

A standard form of partially nonlinear model is defined as

$$Y_i = f(X_i, \delta) + h(U_i) + \varepsilon_i, i = 1, 2, ..., n, \tag{1.1}$$

where $X_i = (x_{i1}, x_{i2}, ..., x_{ip})(p \geq 1)$ and $U_i = (u_{i1}, u_{i2}, ..., u_{iq})(q \geq 1)$ are considered as the vectors of independent and identically distributed explanatory variables, respectively, Y_i is the response variable for the ith case, $\delta =$

Dicle University.

$(\beta_1, \beta_2, ..., \beta_p)^T$ is unknown parameters' vector, $f(.,.)$ is a pre-assigned function, h is an unknown smooth function from o^q to o^l, and $\varepsilon_i (i = 1, 2, ..., n)$ are independent random errors with $E(\varepsilon_i) = 0, Var(\varepsilon_i) = \sigma^2$, respectively.

PNLMs are widely used in the literature due to their usefulness mentioned above. Li and Mei in 2013 [21] developed new estimation procedures for parameters in the parametric component and they formed consistency and asymptotic normality of the estimator which they achieved. Also, they proposed estimation procedures that consider a generalized F test for the nonparametric component in the PNLMs. Severini and Wong in 1992 [30] introduced a geometric framework that contains the concept of the least favorable curve for PNLMs [30]. Zhong et al. in 2000 [37] established three types of developed approximate confidence regions for the parameter in terms of curvatures for PNLMs, considering Severin's geometric framework. Application of the finite series approximation method to a partially nonlinear model and its some new results were handled by Xie et al. in 1997 [13]. Hung and Chen in 2008 [18] studied the parameter estimation problem for the nonlinear partial spline model, $f(X, \delta) = \delta^T X$ that is a special form of PNLMs when a nonparametric component is approximated by some graduating function. Wang and Ke in 2009 [34] developed an estimation problem for smoothing spline semi-parametric nonlinear regression models by considering the penalized likelihood and they solved it by Gauss-Newton and back-fitting algorithms. The parameters in the problem were estimated by employing generalized cross-validation(GCV) [8] and generalized maximum likelihood methods (GML) [36].

This chapter proposes an estimation procedure for PNLMs where its nonparametric component $h(U_i)$ is considered as an additive nonparametric component [6, 32]. To achieve an estimation of parameter for both nonparametric and parametric parts, firstly, we establish a nonlinear least square estimation problem based on the Taylor expansion of nonlinear function $f(.)$ at initial value $\hat{\delta}_c$ where $\hat{\delta}_c$ is a consistent estimate of δ. Secondly, we establish a kernel based bridge problem to estimate the nonparametric component of PNLMs, say $h(.)$. Then, we solve the problem that we established with the famous method of convex optimization called conic quadratic programming (CQP) [5]. Finally, for each δ we evaluate $\hat{\theta}$ and hence, h in kernel based bridge problem that gives a new curve over h. Then, we produce a solution profile for h and choose the $\hat{\delta}$ which is the minimum over all these curves as an estimation of δ.

The remainder of this chapter is formed as follows. The second Section includes an introduction of the bridge estimator for the regression model. Additive approximation and kernel based bridge estimation problem for PNLMs are handled in the third section. The fourth section presents the construction of the estimation problem established in the third section as a CQP problem in order to use the advantages of CQP. Finally, the fifth section gives a short conclusion.

1.2 Bridge Estimators

Penalized estimation methods such as penalized linear least squares and penalized likelihood, have drawn much attention in recent years, and it has been used quite a lot by many researchers because they present a method for selecting of variables and estimating of parameters simultaneously in linear regression given as

$$y_i = x_i^T \delta + \varepsilon_i, i = 1, 2, ..., n. \tag{1.2}$$

Here, $y_i \in o$ is a i-th response variable, $x_i (i = 1, 2, ..., n)$ is a p-vector of covariates, δ is a p-vector of unknown parameters, and ε_i is an n-vector of identically distributed, independent random errors. The bridge estimation method suggested by Frank and Friedman in 1993 [11] consists of a large class of the penalty methods considering penalty function $\sum |\delta_j|^\alpha$ with $\alpha > 0$. Bridge estimator, $\hat{\delta}_B$, can be determined by solving optimization problem given as

$$\text{minimize}_\delta \sum_{i=1}^{n} (y_i - x_i^T \delta)^2 + \varphi \sum_{j=1}^{p} |\delta_j|^\alpha, \tag{1.3}$$

where $|.|$ is the L_2-norm of the vector, φ is a penalty parameter that provides a trade off between the first and the second term. As seen in Equation 1.3, the objective function is penalized by the L_α-norm to obtain bridge estimator $\hat{\delta}_B$ and it shrinks the estimates of the parameters in Equation 1.2 towards 0. Liu et al. in 2007 [22] discussed the effect of the L_α penalty with different cases of α. If $\alpha = 1$, bridge estimation produces Lasso (Least Absolute Shrinkage Operator) [32]; if $\alpha = 2$, it produces Ridge or Tikhonov regularization [17] estimation. For $\alpha \leq 1$, the bridge estimator manages to select significant variables for the regression model by shrinking small $|\delta_j|$s to exact zeros. However, Knight and Fu in 2000 [20] handled the asymptotic distributions of bridge estimators when the number of covariates is fixed and they noted that the amount of shrinkage towards zero increases with the magnitude of the regression coefficients being estimated in case of $\alpha > 1$.

Also, Liu et al. in 2007 [22] pointed out that in penalized estimation problem, to obtain acceptable bias for large parameters, the value of α is not chosen too high than necessary.

They give the following example for $\alpha \in [0, 2]$ to explain this situation.

Liu et al. in 2007 [22] considered a simple linear regression model with one parameter δ and one observation $y = \delta + \varepsilon$, where ε is a random error with mean zero and variance σ^2 for illustrating the effect of L_α penalties with respect to different α. As a result of this example, they made the following inferences:

i) Bridge solution is ridge $\hat{\delta} = y/(\varphi + 1)$ in case $\alpha = 2$ and it is biased with $Var(\hat{\delta}) = Var(y/(\varphi + 1)^2)$. Hence, $\hat{\delta}$ is better than y when the bias is smaller compared to variance deduction.

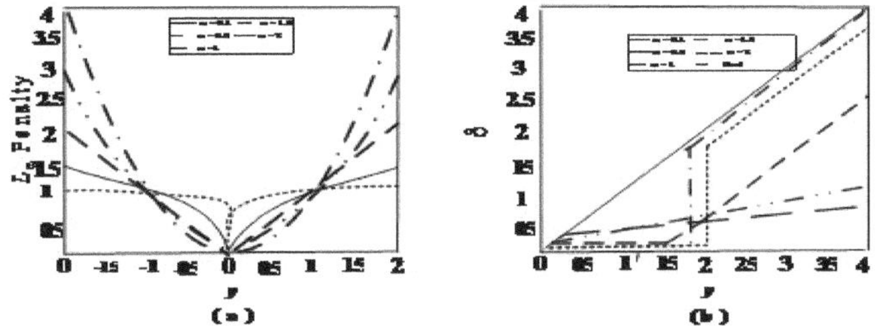

Figure 1.1: **a**: Plots of L_α penalties for different α, **b**: The solutions $\hat{\delta} = \text{argmin} F_\alpha(\delta)$ with respect to the L_α penalties in **(a)** with $\varphi = 3$ where $F_\alpha(\delta) = (\delta - y)^2 + \varphi|\delta|^\alpha$ [22].

ii) Bridge solution is a Lasso solution $\hat{\delta} = sgn(y)[|y| - \varphi/2]$ that gives a thresholding rule, since small $|y|$ leads to a zero solution.

iii) They conclude that $\hat{\delta} = 0 \iff \varphi > |y|^{2-\alpha}\left(\frac{2}{2-\alpha}\right)\left[\frac{2(1-\alpha)}{2-\alpha}\right]^{1-\alpha}$ in case $\alpha \in (0,1)$, that is, $|y| < \left[\varphi\left(\frac{2-\alpha}{2}\right)\left(\frac{2-\alpha}{2(1-\alpha)}\right)^{1-\alpha}\right]^{1/(2-\alpha)}$ [20].

As pointed out in [24], L_α penalty in the problem Equation 1.3 is strictly convex when $\alpha > 1$ and strictly non-convex when $\alpha < 1$. When $\alpha = 1$, it is still convex, but, not differentiable at the origin. It is clearly shown that the elasticnet penalty is between $\alpha = 1$ (Lasso) and $\alpha = 2$ (ridge), and it is strictly convex. As a result, when $\alpha \geq 1$, the problem Equation 1.3 is convex and solvable without using the approximation. Therefore, we handle bridge estimators for partially nonlinear model by using advantages of convex optimization, especially, conic quadratic programming [5].

1.3 PNLMs with Additive Approximation and Bridge Estimation

1.3.1 Construction of additive nonparametric component

In this section, we present the form of the bridge penalty for PNLM. Let us consider $\{(y_i, x_i, u_i), i = 1, 2, ..., n\}$ a random sample from model expressed by Equation 1.4. The nonlinear least squares objective function for Equation 1.4 is written as

$$Q(h, \delta) = \sum_{i=1}^{n} (y_i - f(x_i, \delta - h(u_i))^2. \tag{1.4}$$

Estimation procedure for parameters in the PNLM consists of two-step through Equation 1.4. In the first step, a linear approximation of $f(., \delta)$ is taken into consideration through Taylor expansion since $f(., \delta)$ is nonlinear with respect to δ. Then, we try to find the minimizer of Equation 1.4 by solving normal equations obtained from the derivative of $Q(h,\delta)$ concerning δ. However, normal equations cannot be solved analytically due to their non-linearity; therefore, iterative techniques such as the Newton-Raphson algorithm should be used [3]. By applying Taylor expansion to $f(.,\delta)$, at $\hat{\delta}_c$ where $\hat{\delta}_c$ is a consistent estimate of δ as an initial point, thus, we get

$$f(x,\delta) = f(x,\hat{\delta}_c) + f'(x,\hat{\delta}_c)^T\left(\delta - \hat{\delta}_c\right) + o_p\left(\| \delta - \hat{\delta}_c \|\right). \tag{1.5}$$

The initial point $\hat{\delta}_c$ in Equation 1.5 is obtained from the solution of the following nonlinear least squares optimization problem:

$$\hat{\delta}_c = \text{argmin}_\delta \sum_{i=1}^{n} \left(y_{(i+1)} - y_{(i)} - f(x_{(i+1)},\delta) + f(x_{(i)},\delta)\right)^2, \tag{1.6}$$

where $\left(x_{(i)}, t_{(i)}, y_{(i)}\right)$ $(i = 1,2,...,n)$ is an ordered sample from the smallest to the largest according to the value of the variable u_i [35]. Li and Mei in 2013 [21] have shown that under some conditions, $\hat{\delta}_c$ is root n consistent. Thus, the ith sample for response variable Y_i, y_i can be written as

$$y_i = f(x_i,\hat{\delta}_c) + f'(x_i,\hat{\delta}_c)^T\left(\delta - \hat{\delta}_c\right) + h(u_i) + \varepsilon_i. \tag{1.7}$$

Let $z_i = y_i - f(x_i,\hat{\delta}_c) + f'(x_i,\hat{\delta}_c)^T f(x_i,\hat{\delta}_c)$. Then, we get the following linear approximation model,

$$z_i = f'(x_i,\hat{\delta}_c)^T\delta + h(u_i) + \varepsilon_i \tag{1.8}$$

or in matrix

$$z = F\delta + h(u) + \varepsilon, \tag{1.9}$$

where $z = (z_1, z_2, ..., z_n)^T$ is n-vector of adjusted response variables, F is $(n \times p)$-dimensional derivative matrix whose (i,j) element is $\partial f(x_i,\delta)/\partial\delta_j|_{(\delta=\hat{\delta}_c)}$ and $h(u)$ is n-vector of regression function $h(u_i)$. In the second step, available estimation methods can be directly used to estimate δ by considering the partial linear model given as Equation 1.9.

Here, we prefer the profile least squares technique which is also a two-step process. In the first step, for a given δ let $y_i = z_i - f'(x_i,\hat{\delta}_c)^T\delta$. Then, we rewrite model Equation 1.4 as

$$v = h(u) + \varepsilon, \tag{1.10}$$

where $v = (v_1, v_2, ..., v_n)^T$ can be considered sample from model Equation 1.10 and $h(u) = (h(u_1), h(u_2), ..., h(u_n))^T$ is considered n-vector for unknown regression function. The model Equation 1.10 is a q-dimensional nonparametric regression model. Hence, $h(.)$ may be estimated by one of the nonparametric estimation methods such as smoothing spline [10], k-nearest-neighbors [16], kernel estimation, [29] and local least squares estimation [23] for characterizing nonlinear trend in the model.

The large number of explanatory variables $U_j (j = 1,2,...,q)$ in the model Equation 1.10 causes an increase in the variance of nonparametric estimators and therefore, an increase in test error. This situation, called the curse of dimensionality [4], makes the interpretations of the model very difficult and prevents obtaining reliable results. These challenges of nonparametric regression were overcome with *Additive Models* recommended by Stone in 1985 [32]. Additive models [32] offer estimates that have lower variance than nonparametric models, and can have a lower bias than parametric ones. In additive approximation, the change in the response variable corresponding to each explanatory variable is explained by the estimation of individual terms and it employs univariate smoothing. Therefore, the additive regression model will be considered for estimation of smooth function $h(.)$ to obtain a profile nonlinear least square [21] estimate of δ. Given n realizations for pairs $(u_i, v_i)(i = 1,2,...,n)$ with each $u_i = (u_{i1},...,u_{iq})$, the additive model for $h(u_i)$ is written as

$$E(V_i|u_{i1},...,u_{iq}) = h(u_i) = \delta_0 + \sum_{j=1}^{q} h_j(u_{ij}), i = 1,...,n, \tag{1.11}$$

under the assumption $E(h_j(u_{ij})) = 0$ in order to avoid a different intercept in each h_j function [6, 32]. The additive estimate of each function, $h(u_i)$ by considering Equation 1.11 is obtained by iteration scheme, called back-fitting algorithm, which is proposed by Friedman and Stuetzle in 1981 [14].

In this study for additive approximation of function $h(u)$, the functions h_j will be considered as spline functions, that is, linear combination of the parametrical form

$$h_j(u) = \sum_{l=1}^{d_j} \theta_l^j g_l^j(u), \tag{1.12}$$

where $g_l^j : R \longrightarrow R$ is the l-th transformation (base spline) of $u, (l = 1,2,...,d_j)$, θ_l^j is the (l, j)-th entry of the family

$$\theta = \left(\theta_l^j\right)$$

in which $l = 1,...,d_j$ and $j = 1,...,q$ as well as for the sake of simplicity, by introducing additional terms with coefficients 0, we may assume that $g_l^j \equiv g_l$, $d_j \equiv d, (j = 1,2,...,q)$ such that the family becomes a matrix.

Natural spline [26], B-spline [9] and multivariate adaptive regression spline [12] are examples of spline functions commonly used in data analysis.

Let us now explicitly insert the parametrical form Equation 1.12 of the functions h_j into Equation 1.11. Then, Equation 1.11 looks as follows:

$$E(V_i|u_{i1},...,u_{iq}) = h(u_i) = \delta_0 + \sum_{j=1}^{q}\sum_{l=1}^{d} \theta_l^j g_l(u_{ij}), i = 1,...,n. \tag{1.13}$$

For all $i = 1,2,...,n$ we can write

$$\begin{aligned}\Sigma_{j=1}^{q}\Sigma_{l=1}^{d}\theta_l^j g_l(u_{ij}) &= \theta_1^1 u_1(u_{ij}) + ... + \theta_d^1 u_d(u_{ij}) + ... + \theta_1^q u_1(u_{iq}) + ... + \theta_d^q u_d(u_{iq}) \\ &= (g_1(u_{i1},...,g_d(u_{i1}))\left(\theta_1^1,...,\theta_d^1\right)^T + ... + (g_1(u_{im},...,g_d(u_{im}))\left(\theta_1^m,...,\theta_d^m\right)^T\end{aligned} \tag{1.14}$$

or

$$\sum_{j=1}^{q}\sum_{l=1}^{d} \theta_l^j g_l(u_{ij}) = G_i^1\theta_i^1 + ... + G_i^q\theta_i^q + (G_i^1,...,G_i^q)\left(\theta_i^{1T},...,\theta_i^{qT}\right)^T = G_i\theta \tag{1.15}$$

where $\theta^j := \left(\theta_1^j,...,\theta_d^j\right)^T$, $\theta = \left(\theta^{1T},...,\theta^{qT}\right)$, $G_i^j := (g_1(u_{ij}),...,g_d(u_{ij}))$ and $G_i := \left(G_i^1,...,G_i^m\right)$ $(i = 1,2,...,n)$. If Equation 1.15 is used in Equation 1.11 and assuming that δ_0 is fixed via the estimation $\hat{\delta}_{a0} := ave(i|i = 1,2,...,n)$ by the arithmetic mean of the values i, then, Equation 1.11 is obtained $\hat{\delta}_0$ as

$$h(u_i) = \hat{\delta}_0 + G_i\theta, i = 1,...,n. \tag{1.16}$$

Hence, as a result of this last equation, Equation 1.10 turns into the following form:

$$v = \hat{\delta}_0 1 + G\theta + \varepsilon \tag{1.17}$$

where 1 is a n-vector of ones, G is an $(n \times qd)$-dimensional matrix with ith row $G_i := \left(G_i^1,...,G_i^m\right)$.

1.3.2 Kernel based bridge estimation for PNLMs

To obtain bridge estimation of the parameters θ in Equation 1.17, the objective function is defined as

$$L_B(\hat{\delta}_0, \theta_l^1,...,\theta_d^q) := \sum_{i=1}^{n}\{v_i - \hat{\delta}_0 - G_i\theta\}^T + \varphi\sum_{j=1}^{q}\sum_{l=1}^{d}|\theta_l^j|^\alpha, \tag{1.18}$$

which is just a penalized least square objective function penalized by L_α-norm. This penalization provides shrinking of the estimates of the parameters (θ_l^j) in Equation 1.17 towards 0. Here, $\varphi > 0$ is a penalty or smoothing parameter that provides a trade-off between the goodness of data fitting expressed by the first sum and the penalty function expressed with the second sum. As can be seen, the smoothing parameter φ influences the smoothness of a fitted curve. Therefore, it should be estimated by one of the well known methods such as generalized cross validation (GCV) [8], Akaike information criteria (AIC) [7] and minimization of an unbiased risk estimator (UBRE) [8]. The goal of smoothness is sometimes also called stability, robustness, or regularity. In fact, in the theory of inverse problems one wants to guarantee that the estimation is sufficiently stable with respect to noise and other forms of perturbation.

The bridge estimation of θ, $\hat{\theta}_B$ is obtained by solution of the optimization problem

$$\text{minimize}_{\theta} L\left(\hat{\delta}_0, \theta_l^1, ..., \theta_d^q\right). \tag{1.19}$$

To solve problem Equation 1.18, we consider the kernel estimation method [29] developed for modelling strong non-linearity between independent and dependent variables. This method provides estimates of the regression function by stating the nature of the local neighborhood expressed by a *kernel function* $K_\lambda(x_0, x)$, and the nature of the class of regular functions fitted locally. In this sense, a transformation of the original data is used through kernel functions which are considered as weights, and they form a kernel matrix to produce weighted average estimators. Thus, by using this method, the complexity of the calculations is considerably reduced since the model is performed by considering the kernel matrix that summarizes the similarity in the observation values instead of the original data.

The simplest form of kernel estimate is the Nadaraya-Watson weighted average [29].

$$\hat{h}(u_0) = \frac{\sum_{i=1}^{n} K_\lambda(u_0, u_i) v_i}{\sum_{i=1}^{n} K_\lambda(u_0, u_i)}, \tag{1.20}$$

where $K_\lambda(u_0, u_i) := K(\| u_i - u_0 \|_2 / \lambda$ a kernel function defined as $K : o \longrightarrow o$, providing

$$\int K(u)du = 1, K(-x_0) = K(x_0). \tag{1.21}$$

The most typical kernels functions for $u_i \in o^q$ are (i) Uniform, (ii) Epanechnikov, (iii) Gaussian, Quartic (biweight), and (iv) Tricube (triweight) given as follows:

i $K(u) = \frac{1}{2} 1_{(\|u\| \leq 1)0}$

ii $K(u) = \frac{3}{4}(1 - \| u \|^2) 1_{(1 - \|u\| \leq 1)0}$

iii $K(u) = \frac{15}{16}(1 - \| u \|^2) 1_{(1 - \|u\| \leq 1)0}$

iv $K(u) = \frac{35}{32}(1 - \| u \|^2)^3 1_{(1 - \|u\| \leq 1)0}$

Hence, the bridge estimate of θ in Equation 1.17 based on a kernel function can be obtained by minimizing the penalized residual sum of squares given

$$L_{KB}(\hat{\delta}_0, \theta_l^1, ..., \theta_d^q) := \sum_{i=1}^{n} K_\lambda(u_0, u_i)\{v_i - \hat{\delta}_0 - G_i\theta\}^2 + \varphi \sum_{i=1}^{q} \sum_{l=1}^{d} |\theta_l^j|^\alpha, \tag{1.22}$$

where dimensions of all matrices and all vectors are the same as the corresponding vectors and matrices in Equation 1.22. If we take $b_i = v_i - \hat{\delta}_0 (i = 1, 2, ..., n)$, $b = (b_1, ..., b_n)^T, A = K^{1/2}G, a = K^{1/2}b$ where $K^{1/2}$ is $(n \times n)$-dimensional diagonal matrix with ith diagonal element $[L_\lambda(u_0, u_i)]^{1/2}$, then, kernel based penalized residual sum of squares looks as

$$\begin{aligned} L_{KB}(\hat{\delta}_0, \theta_l^1, ..., \theta_d^q) : &= \sum_{i=1}^n K_\lambda(u_0, u_i)\{v_i - \hat{\delta}_0 - G_i\theta\}^2 + \varphi \sum_{i=1}^q \sum_{l=1}^d, |\theta_l^j|^\alpha. \\ &= \sum_{i=1}^n K_\lambda(u_0, u_i)\{b_i - G_i\theta\}^2 + \varphi \sum_{i=1}^q \sum_{l=1}^d, |\theta_l^j|^\alpha. \\ &= (G\theta - b)^T K_\lambda (G\theta - b) + \varphi \sum_{i=1}^q \sum_{l=1}^d, |\theta_l^j|^\alpha. \\ &= \| A\theta - a \|^2 + \varphi \sum_{i=1}^q \sum_{l=1}^d, |\theta_l^j|^\alpha. \end{aligned} \tag{1.23}$$

The kernel based bridge estimation of θ, $\hat{\theta}_{KB}$ is obtained by solution of the optimization problem

$$\text{minimize}_\theta L(\hat{\delta}_0, \theta_l^1, ..., \theta_d^q). \tag{1.24}$$

We handle two well known special cases of kernel based bridge estimator. When $\alpha = 2$, the kernel based bridge estimator,$\hat{\theta}_{KB}$ will be equivalent to the kernel based ridge estimator $\hat{\theta}_{KB}^R$ [17] and it is obtained from the solution of the following problem:

$$\text{minimize}_\theta \| A\theta - a \|_2^2 + \varphi \| \theta \|_2^2, \tag{1.25}$$

where $\| 0 \|_2$ stands for Euclidean norm.

The optimization problem given in Equation 1.24 can be solved by using SVD of coefficient matrix A [2] and its solution is

$$\hat{\theta}_{KB}^R = \left(A^T A + \varphi\right)^{-1} A^T a. \tag{1.26}$$

For a suitable value of φ, the ridge estimator has a smaller mean squared error than that of the least squares estimator. When $\alpha = 1$, the kernel based bridge estimator, $\hat{\theta}_{KB}$ will be equivalent to the kernel based Lasso estimator, $\hat{\theta}_{KB}^{Lasso}$ that is the solution of the following problem:

$$\text{minimize}_\theta \| A\theta - a \|_2^2 + \varphi \| \theta \|_1 \tag{1.27}$$

where $\| \theta \|_1 := \sum_{j=1}^q \sum_{l=1}^d |\theta_l^j|^\alpha$. If some components of T are 0, the objective function in problem Equation 1.27 will be non-differentiable, so the problem cannot be solved by standard unconstrained optimization methods.

In this section, we consider a convex optimization method called conic quadratic programming (CQP) to solve problem Equation 1.22, as they provide computationally easy study and theoretically efficient solutions, and this solution will be named C_K-bridge estimation.

1.4 On Conic Optimization and Its Application to Kernel Based Bridge Problem

1.4.1 *Convex and conic optimization*

Convex optimization [5] is a special class of mathematical optimization problems such as least-squares and linear programming, and it handles problems aiming to minimize a convex function based on a convex set. The great advantage of expressing a problem as a convex optimization problem is to obtain a reliable and effective solution employing interior-point methods. Convex optimization programs have been employed for many years in scientific research including both theory and practice, as they present a strong theory of duality that has a very interesting comment in terms of the original problem. It has also found wide application in combinatorial optimization and global optimization, as it is capable of finding optimal value bounds as well as its approximate solution for optimization problems. Convex optimization contains different important classes of optimization problems such as CQP which is considered for our problem, semidefinite programming, and geometric programming. Brief information related with CQP will be given in the following by benefiting from [5].

A CQP is a conic problem

$$\text{minimize}_{\phi} c^T \phi, \quad \text{where} \quad S\phi - s \in C. \tag{1.28}$$

Here, the cone C consists of direct product of *"Lorentz cones"* defined as

$$L^{n_i+1} = \{\phi = (\phi_1, \phi_2, ..., \phi_{n+1})^T \in R^{n_i+1} | \phi_{n_i+1} \geq \sqrt{\phi_1^2 + \phi_2^2 + ... + \phi_{n_i}^2}\}(n_i \geq 1, n_i \in Y). \tag{1.29}$$

The geometric interpretation of a quadratic (or second-order) cone is shown in Figure 1.2 for a cone with n_i variables, and illustrates how the boundary of the cone resembles an ice-cream cone. The 1-dimensional quadratic cone simply states non-negativity $\phi_{n_i+1} \geq 0$. More generally, partitioning the data matrix $[S_i; s_i]$ by

$$[S_i; s_i] = \begin{bmatrix} D_i & d_i \\ p_i^T & q_i \end{bmatrix},$$

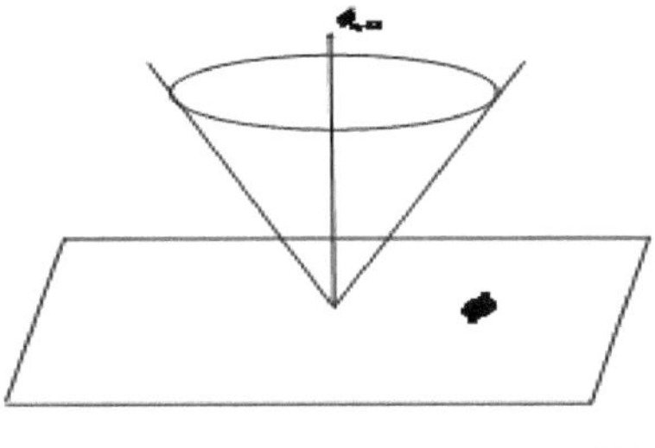

Figure 1.2: Boundary of quadratic cone $\phi_{n_i+1} \geq \sqrt{\phi_1^2 + \phi_2^2 + ... + \phi_{n_i}^2}$.

then, its standard form may be written as

$$\underset{\alpha}{\text{minimize}}\ \phi \text{ such that } \| D_i\phi - d_i \|_2 \leq p_i^T \phi - q_i (i = 0, 2, ..., k). \tag{1.30}$$

The dual problem for problem Equation 1.30 is

$$\underset{\Psi}{\text{maximize}}\ s^T\Psi, \text{ where } s^T\Psi = c, \Psi \in C. \tag{1.31}$$

If we write Ψ as $\Psi := (\Psi_1^T, \Psi_2^T, ..., \Psi_k^T)^T$ with m_i-dimensional blocks Ψ_i, then the dual problem can be stated as follows:

$$\underset{\Psi_1,...,\Psi_n}{\text{maximize}} \sum_{i=1^n} s_i^T\Psi_i, \text{ where } \sum_{i=1}^{k} S_i^T\Psi_i = c \text{ and } \Psi_i \in L^{n_i} (i = 1, 2, , ..., k). \tag{1.32}$$

If it is taken $\Psi_i = (\kappa_i^T d_i + \zeta_i)^T$ with a scalar component ζ_i and using the meaning of "$\geq_{L^n} 0$", it can be shown that following form is the problem dual of problem Equation 1.30

$$\text{maximize}_{(\kappa_i),(\zeta_i)} \sum_{i=1}^{k} [\kappa_i^T d_i + \zeta_i q_i], \text{where} \sum_{i=1}^{k} [D_i^T \kappa_i + \zeta_i p_i] = c, \| \kappa_i \|_2 \leq \zeta_i (i = 1, 2, ..., k). \tag{1.33}$$

The design variables in Equation 1.33 are column vectors κ_i, having the same dimensions as the vectors d_i, and reals $\zeta_i (i = 1, 2, ..., k)$. The problems in Equation 1.30 and Equation 1.33 are standard forms of a conic quadratic problem and of its dual, respectively.

1.4.2 Conic kernel based bridge estimator (C-KBBE)

We want to solve problem Equation 1.24 by CQP to benefit the advantage of convex optimization mentioned above. Therefore, the problem Equation 1.24 should be written as a standard form of CQP which is a well structured convex optimization problem. For this reason we will consider two special cases of the kernel based bridge problem, Equation 1.24 given in Equation 1.25 and Equation 1.27.

1.4.2.1 C-KBBE for case $(\alpha = 2)$

Let firstly tackle kernel based ridge problem in Equation 1.25 for obtaining conic kernel based bridge estimator of θ, $\hat{\theta}_{CKBB}^R$. The problem in Equation 1.25 can be formulated as a CQP problem based on an appropriate selection of bound T. Thus, the problem in Equation 1.25 can be written as follow:

$$\text{minimize}_\theta \quad \| A\theta - a \|_2^2 \text{ subject to } \| \theta \|_2^2 \leq T. \tag{1.34}$$

The biggest contribution of CQP to the solution of the problem in Equation 1.33 is that the smoothing parameter φ does not need to be calculated separately. In other words, the solution of the problem in Equation 1.33 by CQP creates an alternative solution to determine φ. In this sense, bound T should be found as a result of a careful learning process, with the help of model-free or model-based methods [16]. The problem in Equation 1.34 involves the least-squares objective function $\| A\theta - a \|_2^2$ and the inequality constraint function $- \| \theta \|_2^2 + T$ that should be non-negative for feasibility. Then, optimization problem in Equation 1.34 is equivalently written as follows:

$$\text{minimize}_{t,\theta} \text{subject to } \| A\theta - a \|_2^{2\;2}, t \geq 0 \text{ subject } \| \theta \|_2^2 \leq T \tag{1.35}$$

or, equivalently again

$$\text{minimize}_{t,\theta} \text{subject to } \| A\theta - a \|_2^2 \leq t^2, t \geq 0 \, \| \theta \|_2 \leq \sqrt{T}. \tag{1.36}$$

If optimization problem in Equation 1.36 is compared with the standard form of CQP, it is observed that it is a CQP programme with

$$c = \left(1, 0_{qd}^T\right)^T, \phi = \left(t, \theta^T\right)^T, D_1 = \left(0_n, A\right), d_1 = a, p_1 = (1, 0, \ldots, 0)^T, q_1 = 0$$

$$D_2 = \left(0_{qd}, I_n, d_2 = 0_{qd}, p_2 = 0_{qd+1}, \text{and } q_2 = -\sqrt{T}.\right.$$

The *dual problem* of the problem Equation 1.35 is written

$$\begin{aligned} &\text{mazimize} \left(a^T, 0\right) \omega_1 + \left(0_{qd}^T, -\sqrt{T}\right) \omega_2 \\ &\text{such that} \begin{bmatrix} 0_n^T & 1 \\ A_{n\times qd}^T & 0_{qd} \end{bmatrix} \omega_1 + \begin{bmatrix} 0_{qd}^T & 0 \\ I_{qd\times qd} & o_{qd} \end{bmatrix} \omega_2 = \begin{bmatrix} 1 \\ 0_{qd} \end{bmatrix}, \\ &\omega_1 \in L^{n+1}, \omega_2 \in L^{qd+1}. \end{aligned} \tag{1.37}$$

In order to write the equations regarding the optimality conditions of the problem in Equation 1.36, it is reformulated as follows:

$$\begin{aligned} &\text{minimize}_{t,\theta} \quad t, \\ &\text{such that} \quad \chi = \begin{bmatrix} 0_n & A_{n\times qd} \\ 1 & 0_{qd}^T \end{bmatrix} \begin{bmatrix} t \\ \theta \end{bmatrix} + \begin{bmatrix} -a \\ 0 \end{bmatrix} \\ &\qquad\qquad \eta = \begin{bmatrix} 0_n & I \\ 0 & 0_{qd}^T \end{bmatrix} \begin{bmatrix} t \\ \theta \end{bmatrix} + \begin{bmatrix} 0_{qd} \\ \sqrt{T} \end{bmatrix}, \\ &\qquad\qquad \chi \in L^{n+1}, \eta \in L^{qd+1} \end{aligned} \tag{1.38}$$

where L^{n+1}, L^{qd+1} are the $(n+1)$ and $(qd+1)$-dimensional *ice-cream* (or *second-order*, or *Lorentz*) *cones*.

Thus, primal dual optimal solutions $(t, \theta, \chi, \eta, \omega_1, \omega_2)$ are obtained if and only if the following constrains are provided:

$$\begin{aligned}
&\chi = \begin{bmatrix} 0_n & A_{n\times qd} \\ 1 & 0_{qd}^T \end{bmatrix} \begin{bmatrix} t \\ \theta \end{bmatrix} + \begin{bmatrix} -a \\ 0 \end{bmatrix}. \\
&\eta = \begin{bmatrix} 0_{qd} & I_{qd\times qd} \\ 0 & 0_{qd}^T \end{bmatrix} \begin{bmatrix} t \\ \theta_{qd} \end{bmatrix} + \begin{bmatrix} 0_{qd} \\ \sqrt{T} \end{bmatrix}. \\
&\begin{bmatrix} 0_n^T & 1 \\ A_{n\times qd}^T & 0_{qd} \end{bmatrix} \omega_1 + \begin{bmatrix} 0_{qd}^T & 0 \\ I_{qd\times qd} & o_{qd} \end{bmatrix} \omega_2 = \begin{bmatrix} 1 \\ 0_{qd} \end{bmatrix}. \\
&\omega_1^T \chi = 0, \omega_2^T \eta = 0, \\
&\omega_1 \in L^{n+1}, \omega_2 \in L^{qd+1}, \chi \in L^{n+1}, \eta \in L^{qd+1}.
\end{aligned} \tag{1.39}$$

1.4.2.2 C-KBBE for case $(\alpha = 1)$

As stated before, the kernel based bridge estimator is equivalent to the kernel based Lasso in case $(\alpha = 1)$ and it can be found by the solution of problem in Equation 1.27, that is, a non-smooth optimization problem. The most common methods used to solve such problems are Quadratic Programming, iterated ridge regression [17] and methods that use sub-gradient strategies [31]. Quadratic Programming needs 2^{qd} constraint functions for solving the problem in Equation 1.27. Unfortunately, this situation makes it very difficult to perform iterations over all the constraints generated by this expansion for non-trivial values of rp. The methods involving sub-gradient strategies are very difficult for large scale problems, as they may need so many coordinate updates that they become unpractical. Therefore, we want to use conic optimization for solving problem in Equation 1.27 to avoid the disadvantages of mentioned two methods. However, the problem cannot be written as a CQP problem since the objective function is nondifferentiable in L_1-regularization given in Equation 1.27. So, we cannot find conic kernel based bridge estimator of θ, say $\hat{\theta}_{CKBB}^{Lasso}$. For this reason, it should be considered a differentiable approximation method to L_1 -regularization. The *iterated ridge regression* (IRR) [17] method provides a differentiable approximation to L_1-regularization that contains a non-differentiable objective function and it updates multiple variables at each iteration. Therefore, we consider the IRR to L_1-regularization in Equation 1.27 and we will solve it by CQP. IRR method is based on the following approximation:

$$|\theta_l^j| \approx \frac{\theta_l^{j2}}{|\theta_l^{jk}|}, \tag{1.40}$$

where θ_l^{jk} is the value from the previous iteration k.

Substituting this approximation into the unconstrained formulation in Equation 1.27, we can obtain an expression similar to the least-squares estimation with an L_2-penalty (ridge regression) as follows:

$$\text{minimize}_\theta \quad \| A\theta - a \|_2^2 + \varphi \theta^T R \theta, \tag{1.41}$$

where R is $(qd \times qd)$-dimensional diagonal matrix whose diagonal element are consist of $|\theta_l^{jk}|^{-1}$, that is, $R = diag\left(|\theta_1^{lk}|^{-1}, ..., |\theta_d^{lk}|^{-1}, ..., |\theta_1^{qk}|^{-1}, ..., |\theta_d^{qk}|^{-1}\right) := diag\left(|\theta^k|^{-1}\right)$.

The solution of the problem in Equation 1.41 for the kth step is

$$\left(\hat{\theta}_{KBB}^{Lasso}\right)^{k+1} \cong \left(A^T A + \varphi diag|\theta^k|^{-1}\right)^{-1} A^T a. \tag{1.42}$$

However, we should note that this approximation will be numerically unstable when one of the θ_l^{jk} approaches 0. To avoid this problem, we can use a generalized inverse [1, 25] of $|\theta^k|$. Generalized inverse removes values that are too close to 0 from the estimation and avoids this problem. However, this inverse introduces a new problem; variables that are set to 0 can never move away from 0, and, thus, it could potentially lead to sub-optimal results if the initialization is inadequate.

Let us use the Cholesky decomposition of $R = M^T M$ and substitute it into Equation 1.41; thus we get Equation 1.41 as

$$\text{minimize}_\theta \parallel A\theta - a \parallel_2^2 + \varphi\theta \parallel_2^2 . \tag{1.43}$$

As a result of using the approximation in Equation 1.37 in the problem in Equation 1.27, it can easily be noticed that the problem is a specific ridge problem which contains a differentiable objective problem. Hence, it can be expressed as the standard form of CQP problem

$$\begin{array}{ll} \text{minimize}_{t_1,\theta} & t_1 \\ \text{subject to} & \parallel A\theta - a \parallel_2^2 \leq t_1^2, t_1 \geq 0, \parallel M\theta \parallel_2^2 \leq T_1 \end{array} \tag{1.44}$$

or

$$\begin{array}{ll} \text{minimize}_{t_1,\theta} & t_1 \\ \text{subject to} & \parallel A\theta - a \parallel_2 \leq t_1, t_1 \geq 0, \parallel M\theta \parallel_2 \leq \sqrt{T_1} \end{array} \tag{1.45}$$

with $c = \left(1, o_{qd}^T\right)^T$, $\phi = \left(t_1, \theta^T\right)^T$, $D_1 = (0_n, A)$, $d_1 = a$, $p_1 = (1, 0, ..., 0)^T$, $q_1 = 0$, $D_2 = \left(0_{qd}, M\right.$, $d_2 = 0_{qd}$, $p_2 = 0_{qd+1}$, and $q_2 = -\sqrt{T_1}$.

The conic kernel based bridge estimator of θ, $\hat{\theta}_{CKBB}^{Lasso}$ is obtained by solving the problem Equation 1.45. The *dual problem* for problem in Equation 1.45 according to Equation 1.33 is written as

$$\begin{array}{l} \text{maximize}(a^T, 0)\tau_1 + (0_{qd}^T, -\sqrt{T_2})\tau_2 \\ \text{subject to} \begin{bmatrix} 0_n^T & 1 \\ A^T & 0_{qd+1} \end{bmatrix} \tau_1 + \begin{bmatrix} 0_{qd}^T & 0 \\ M^T & 0_{qd} \end{bmatrix} \tau_2 = \begin{bmatrix} 1 \\ 0_{qd+1} \end{bmatrix}. \\ \tau_1 \in L^{n+1}, \tau_2 \in L^{qd+1}. \end{array} \tag{1.46}$$

Moreover, $(t_2, \theta, \chi_2, \eta_2, \tau_1, \tau_2)$ is a primal-dual optimal CKBB-Lasso solution if and only if the following constraints are satisfied:

$$\begin{aligned}
&\chi_2 = \begin{bmatrix} 0_n & A \\ 1 & 0_{qd+1}^T \end{bmatrix} \begin{bmatrix} t_2 \\ \theta \end{bmatrix} + \begin{bmatrix} -a \\ 0 \end{bmatrix}. \\
&\eta_2 = \begin{bmatrix} 0_{qd} & M \\ 0 & 0_{qd+1}^T \end{bmatrix} \begin{bmatrix} t_2 \\ \theta \end{bmatrix} + \begin{bmatrix} 0_{qd} \\ \sqrt{T_2} \end{bmatrix}. \\
&\begin{bmatrix} 0_n^T & 1 \\ A^T & 0_{qd+1} \end{bmatrix} \tau_1 + \begin{bmatrix} 0_{qd}^T & 0 \\ M^T & 0_{qd} \end{bmatrix} \tau_2 = \begin{bmatrix} 1 \\ 0_{qd+1} \end{bmatrix}. \\
&\tau_1^T \chi_2 = 0, \tau_2^T \eta_2 = 0. \\
&\tau_1 \in L^{n+1}, \tau_2 \in L^{qd+1}, \chi_2 \in L^{n+1}, \eta_2 \in L^{qd+1}.
\end{aligned} \tag{1.47}$$

A groundbreaking in solving optimization problems and a widely used method, named interior points [19, 27], are used in order to solve CQP that is a well-structured convex optimization problem.

Let us denote the conic kernel based bridge estimator of θ determined as the solutions of the problems in Equation 1.35 and Equation 1.44 by $\hat{\theta}_{CKBB}$. Hence, the estimation in matrix notation for the regression function can be written as follows if $\hat{\theta}_{CKBB}$ is used in Equation 1.13:

$$\hat{h}_{CKBB} = G\hat{\theta}_{CKBB}. \tag{1.48}$$

As it is known, the estimation of $\hat{\theta}_{CKBB}$ was computed for the initial value of $\delta = \hat{\delta}_c$. So $\hat{\theta}_{CKBB}$ is just a local estimate based on the spline function. If we substitute $\hat{h}_{CKBB}$ for h in Equation 1.9, we get the following equation:

$$w = F\delta + \varepsilon, \tag{1.49}$$

where $w = (w_1, w_2, ..., w_n)^T$ whose ith element is $w_i = z_i - h_{CKB}(u_i)$.

By considering the ordinary linear least square estimation method, the profile conic kernel based bridge estimator of δ, $\hat{\delta}_{CKBB}$ can be written as

$$\begin{aligned}
\hat{\delta}_{CKBB} &= (F^T F)^{-1} F^T \omega. \\
&= (F^T F)^{-1} F^T (z - \hat{h}_{CKBB}(u)). \\
&= (F^T F)^{-1} F^T (z - G\hat{\theta}_{CKBB}).
\end{aligned} \tag{1.50}$$

Then, for each $\delta = \hat{\delta}_{CKBB}$, we evaluate $\hat{\delta}_{CKBB}$ and hence, $\hat{h}_{CKBB}$ in $L_{KB}(\hat{\delta}_0, \theta_l^1, ..., \theta_d^q)$ that gives a new curve over h. In this way, a profile of the solution curve is drawn. Finally, we choose the $\hat{\delta}_{CKBB}$ which is the minimum over all these curves as an estimation of δ.

1.5 Conclusion

This chapter aimed to contribute to the PNLMs, which are widely used in the literature, by combining the flexibility of nonparametric regression and the advantages of continuous optimization. The parametric part of PNLM is represented by nonlinear regression based on the Newton-Raphson algorithm and its nonparametric part is represented by the splines function. Different from existing estimation techniques, the parameter estimation for PNLMs is based on a additive model and a convex optimization problem, called CQP. The present procedure provides an efficient algorithm for complex data structure and reasonable complexity. Hence, a bridge has been built between PNLMs and the powerful tools prepared for well-structured convex optimization problems.

References

[1] P.W. Aitchison. Generalized inverse matrices and their applications. *International Journal of Mathematical Education in Science and Technology*, 13(1): 99–109, 1982.

[2] R.C. Aster, B. Borchers and C.H. Thurber. *Parameter Estimation and Inverse Problems*. Elsevier, 2018.

[3] D.M. Bates and D.G. Watts. Nonlinear Regression Analysis and its Applications. *Wiley, New York*, 1988.

[4] R. Bellman. Adaptive control processes: A Guided Tour Princeton University Press. *Princeton, New Jersey*, USA, p. 96, 1961.

[5] A. Ben-Tal and A. Nemirovski. *Lectures on Modern Convex Optimization: Analysis, Algorithms, and Engineering Applications*. SIAM, 2001.

[6] A. Buja, T. Hastie and R. Tibshirani. Linear smoothers and additive models. *The Annals of Statistics*, pp. 453–510, 1989.

[7] K.P. Burnham and D.R. Anderson. A practical information theoretic approach. *Model Selection and Multimodel Inference*, 2, 2002.

[8] P. Craven and G. Wahba. Smoothing noisy data with spline functions. *Numerische Mathematik*, 31(4): 377–403, 1978.

[9] C. De Boor. *A Practical Guide to Splines Springer-Verlag. New York*, 1978.

[10] J. Fox. *Robust Regression: Appendix to an r and s-plus Companion to Applied Regression*. 2002.

[11] L.E. Frank and J.L. Friedman. A statistical view of some chemometrics regression tools. *Technometrics*, 35(2): 109–135, 1993.

[12] J.H. Friedman. Multivariate adaptive regression splines. *The Annals of Statistics*, pp. 1–67, 1991.

[13] J.H. Xie, M.W. Gao, Z.Q. Liang, Q.Y. Shu, X.Y. Cheng and Q.Z. Xue. The effect of cool-pretreatment on the isolated microspore culture and the free amino acid change of anthers in Japonica rice (*Oryza sativa* L). *Journal of Plant Physiology*, 151(1): 79–82, 1997.

[14] J.H. Friedman and W. Stuetzle. Projection pursuit regression. *Journal of the American statistical Association,* 76(376): 817–823, 1981.

[15] W. Härdle, H. Liang and J. Gao. *Partially Linear Models*. Springer Science & Business Media, 2012.

[16] T. Hastie. R. Tibshirani and J.H. Friedman. *The Elements of Statistical Learning*, 2001.

[17] A.E. Hoerl and R.W. Kennard. Ridge regression: Biased estimation for nonorthogonal problems. *Technometrics*, 12(1): 55–67, 1970.

[18] G.B. Huang and L. Chen. Enhanced random search based incremental extreme learning machine. *Neurocomputing*, 71(16-18): 3460–3468, 2008.

[19] N. Karmarkar. A new polynomial-time algorithm for linear programming. In *Proceedings of the Sixteenth Annual ACM Symposium on Theory of Computing*, pp. 302–311, 1984.

[20] K. Knight and W. Fu. Asymptotics for lasso-type estimators. *Annals of Statistics*, pp. 1356–1378, 2000.

[21] T. Li and C. Mei. Estimation and inference for varying coefficient partially nonlinear models. *Journal of Statistical Planning and Inference*, 143(11): 2023–2037, 2013.

[22] Y. Liu, H.H. Zhang, C. Park and J. Ahn. Support vector machines with adaptive lq penalty. *Computational Statistics & Data Analysis*, 51(12): 6380–6394, 2007.

[23] H.-G. Müller. *Nonparametric Regression Analysis of Longitudinal Data*, volume 46. Springer Science & Business Media, 2012.

[24] C. Park and Y.J. Yoon. Bridge regression: Adaptivity and group selection. *Journal of Statistical Planning and Inference*, 141(11): 3506–3519, 2011.

[25] R.M. Pringle. Generalized inverse matrices with applications to statistics. Technical report, 1971.

[26] S.S. Ray. *Numerical Analysis with Algorithms and Programming.* Chapman and Hall/CRC, 2018.

[27] J. Renegar. *A Mathematical View of Interior-point Methods in Convex Optimization.* SIAM, 2001.

[28] D. Ruppert, M.P. Wand and R.J. Carroll. *Semiparametric Regression.* Number 12. Cambridge University Press, 2003.

[29] B. Schölkopf, A.J. Smola and K.-R. Müller. *Kernel Principal Component Analysis. Advances in Kernel Methods: Support Vector Learning.* Cambridge, MA: MIT Press, pp. 196–201, 1999.

[30] T.A. Severini and W.H. Wong. Profile likelihood and conditionally parametric models. *The Annals of Statistics*, pp. 1768–1802, 1992.

[31] N.Z. Shor. *Minimization Methods for Non-Differentiable Functions.* Springer- Verlag, Berlin, 1985.

[32] C.J. Stone. Additive regression and other nonparametric models. *The Annals of Statistics*, pp. 689–705, 1985.

[33] R. Tibshirani. Regression shrinkage and selection via the lasso. Journal *of the Royal Statistical Society: Series B (Methodological)*, 58(1): 267–288, 1996.

[34] Y. Wang and C. Ke. Smoothing spline semiparametric nonlinear regression models. *Journal of Computational and Graphical Statistics*, 18(1): 165–183, 2009.

[35] A. Yatchew. An elementary estimator of the partial linear model. *Economics Letters*, 57(2): 135–143, 1997.

[36] C.-H. Zhang. Generalized maximum likelihood estimation of normal mixture densities. *Statistica Sinica*, pp. 1297–1318, 2009.

[37] Y.G. Zhong, G.J. Zhang, L. Yang and Y.Z. Zheng. Effects of photoinduced membrane rigidification on the lysosomal permeability to potassium ions. *Photochemistry and Photobiology*, 71(5): 627–633, 2000.

Chapter 2

A Glimpse on the Contributions and Challenges towards more Environmentally-friendly Road Traffic

*Eloisa Macedo** and *Jorge M. Bandeira*

2.1 Introduction

Despite the increasing importance road transport system plays in the society, it also poses some negative externalities, such as accidents, congestion, and inevitably, environmental and noise pollution, with adverse effects on human health. In fact, the transport sector has been responsible for almost 30% of the total EU-28 greenhouse gas (GHG) emissions, in which urban mobility plays a key role, accounting for 40% of all carbon dioxide (CO2) emissions of road transport and approximately 70% of other pollutants [22]. The relevance given to CO2 because it is a GHG that contributes to global warming effects, making it closely associated with climate change. All these issues are mostly related to vehicle technology and driving behavior. Although electric vehicles are expected

Centre for Mechanical Technology and Automation, Dept. Mechanical Engineering, University of Aveiro, Portugal.

* Corresponding author: macedo@ua.pt

to reduce pollutant emissions, the widespread of electric vehicles requires an efficient e-charging infrastructure globally, and forecasts suggest only 30% of the total EU passenger vehicle fleet in 2040 will be electric [22]. This means that there will be a long period in which transport is still powered by internal combustion engines using petroleum-based fuels [19]. It is widely recognized that significant developments have been made to improve road transport efficiency, from manufacturers' side to software developers that demand ever efficient solutions to improve citizens' traveling and quality of life, which environmental consequences should not be neglected.

Managing a complex system such as a road transport network is not easy and requires efficient operational research (OR) procedures. In this context, modelling and simulation play key roles in planning, managing, and anticipating impacts by testing different scenarios. The rapid evolution of vehicle technologies such as automated and connectivity functions will reshape road transport as we know [10, 21]. Although connected and automated vehicles (CAVs) are not fully deployed yet on the roads (only pilot cases are running in some major cities around the world), there will be a long period of CAVs sharing roads with conventional vehicles (CVs). Therefore, it is interesting and important from different perspectives (planning, strategic, operational,...) to anticipate its potential impacts by assuming such technologies are available to the mass market [10] and study various possible scenarios to understand which investments should be prioritized. For the sake of clarity, here CAV functionalities include V2V and V2I communications [18]. Whilst the main motivation for introducing CAVs is to improve traffic efficiency (in terms of congestion), safety, and fuel consumption, significant impacts at other levels can be expected such as in terms of comfort, noise and pollutant emissions, and air quality. The majority of the studies have been conducted to explore the impact of the increase in the market-penetration rate of CAV technologies on the transport system in terms of traffic efficiency (e.g., [20]), traffic flow stability (e.g., [38]), capacity (e.g., [28]), travel time (e.g., [40]), energy efficiency (e.g., [6]) and safety (e.g., [1]). Studies are mostly focused on evaluating the impacts of the deployment of vehicles with automatic cruise control (ACC) and connected ACC (CACC) [25, 33]. In fact, the automated driving systems are expected to affect the vehicle operational characteristics (reduced headways; smooth accelerations/decelerations, longitudinal and lateral behaviors; or smaller gap acceptance thresholds) and can translate into energy and emission benefits [27, 33, 36]. However, energy and environmental impacts are uncertain (for instance, emission levels can be reduced [36] or increased [42]) with CAVs introduction and it is crucial to take an adaptive approach to automation and investigate it under various scenarios [12]. It is thus relevant to rethink road transport systems to support policies to better accommodate such new technologies to minimize negative consequences and maximize their benefits [22]. Therefore, research on mixed traffic environment that impacts at the operational and planning level is needed, as well as studies exploring the impacts of such a new reality to better understand how to best intervene.

In this context, OR opens up a plethora of tools that allow the solution to complex network problems and plan for future challenges in which advanced modeling and simulation are essential for planning, managing and anticipating impacts by testing different scenarios. Establishing scenarios under simulation and exploring consequent driving behaviors is an interesting lens for examining the impacts of CAVs in a mixed traffic network. Thus, a main focus of this chapter is to review how academics working and studying emerging technologies in traffic engineering have been addressing the environmental impacts of CAVs introduction through an OR perspective and to highlight the potential opportunities such new technologies can bring to improve the road traffic ecosystem, mostly regarding the environment and air quality. Given the multitude of approaches on studying many traffic-related impacts of CAVs introduction, the truth is that just very recently, some have been devoting their attention to the environmental consequences/benefits. In particular, the following central research question:

- Based on the extant literature, what environmental and air quality impacts can be expected with the ever-increasing level of connectivity and automated in-vehicle functions in a mixed traffic?

In this realm, the above research question will intrinsically involve other questions, such as:

- Given the fact that CAVs and CVs will share the road for some time, how do CVs and CAVs interact with each other?
- What potential environmental impacts can be expected?
- How can we establish a foundation to better understand and accommodate CAVs in the future?

The main contributions of this chapter are twofold. First, it addresses an important research gap by synthesizing the extant literature on quantifying CAVs introduction environmental impacts, providing a literature snapshot on the current state of research of applications of OR tools (mostly based on simulation and controller approaches) for exploring and anticipating the potential environmental impacts in a mixed road traffic context, where emerging technologies in the form of CAVs will share the roads with CVs. The present survey will focus on literature with a clear objective on quantifying pollutant emissions impacts with CAVs introduction, especially, related to hydrocarbons (HC), carbon monoxide (CO), carbon dioxide (CO2), nitrogen oxides (NOx) and particulate matter (PM). And second, some gaps in the literature are identified and future research directions on this topic are proposed.

2.2 Research Methodology and Descriptive Analysis

This review follows a systematic literature review (SLR) framework to understand which environmental impacts can be expected with CAVs introduction. The first step involves the identification of potentially relevant papers focusing on environmental impacts of CAVs under a mixed traffic context based on OR-related approaches. The documents were selected from the SCOPUS database, the largest database of peer-reviewed literature. The search criteria and key terms were defined, yielding the research query under the fields "title, abstract, keywords" as: (road Operation Research mixed traffic OR conventional vehicle OR human-driven vehicle) AND (automated vehicle OR automated car OR autonomous vehicle OR autonomous car) AND (emissions OR air quality OR environmental impacts OR environment effects OR air pollution OR noise) AND (model OR numerical modelling OR numerical modeling OR simulation OR heuristic OR optimization OR optimization OR multi-objective OR multi-objective OR multi-criteria). The search was conducted in January 2021 for journal papers published over the last five years (between 2015 and 2020). Only papers published in peer-reviewed scientific journals and written in English, and strongly related to the research question were considered. This search resulted in 156 documents, which were screened in terms of the content based on reading the abstracts. In this second step, it was found that the majority of the studies are based on investigating impacts on traffic efficiency and fuel consumption, and were considered outside the scope. Thus, only 31 were related to the evaluation and quantification of the environmental impacts of CAVs introduction in the form of noise and pollutant emissions and air quality. The final step involves reporting the review and analyzing the selected papers.

The VOSviewer software [41] was used to validate the final sample selection by highlighting the frequency of a specific keyword in both samples. For that purpose, a keyword bibliometric analysis based on co-occurrence data was performed. The results on the co-occurrence maps of keywords are illustrated in Figures 2.1 and 2.2 and reveal that the papers included in the selected sample present the core keywords of the objective of this review.

Based on the sample of 31 selected documents, 74% are articles, while 23% are conference papers. Results show that there is an upward trend in the number of papers published in the period considered in this study (2019–35% and 2020–42%), so, there has been increasing attention on evaluating CAVs-related impacts in terms of environment. The three most cited papers on the topic are [16, 35] and [23], with 88, 26, 22 citations, respectively. The Transportation Research Part D: Transport and Environment and Transportation Research Part C: Emerging Technologies represent the two main journals that published papers from the selected sample (4 and 2, respectively). Considering the distribution of publications according to the followed modeling methodology, it was found that

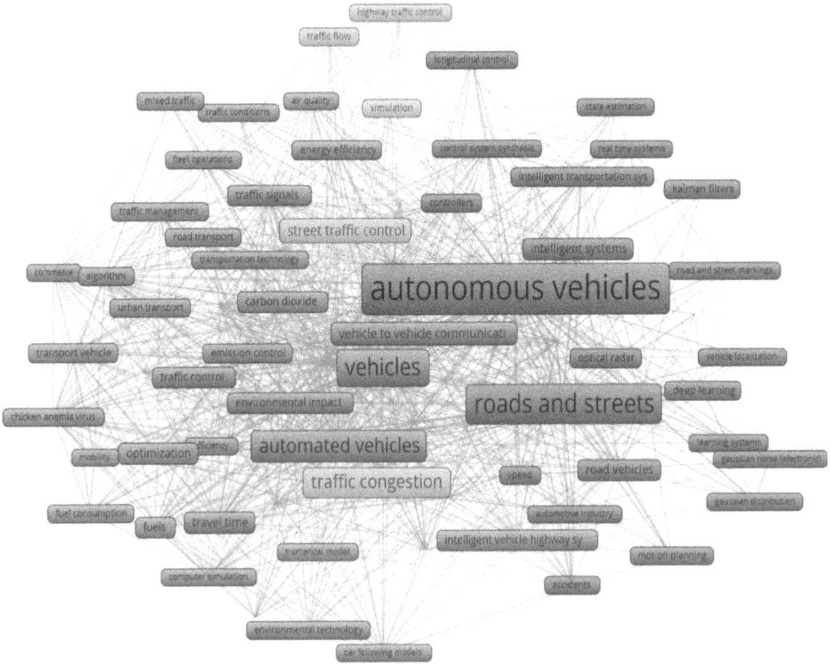

Figure 2.1: Co-occurrence maps for author and indexed keywords on the sample with 156 documents.

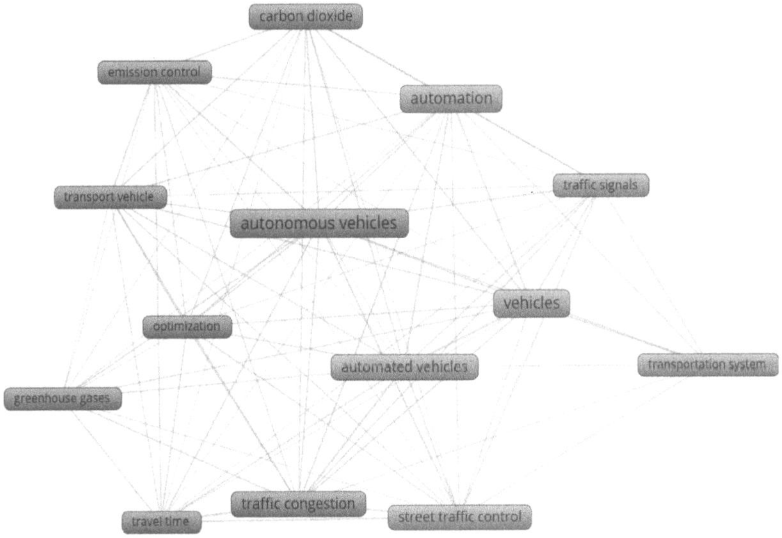

Figure 2.2: Co-occurrence maps for author and indexed keywords for the selected sample.

most of the studies involved simulation modeling (77%) and control strategies (48%) and 32% focused on optimization models.

2.3 Exploring CAVs Environmental Impacts

While conducting this review, we found that the selected papers could be characterized into two main categories for assessing environmental impacts, namely, Eco-routing and Control Strategies and the Model adjustment framework.

2.3.1 Eco-routing and control strategies

In [3], an intersection management algorithm for controlling CAVs was proposed based on a time optimal control problem. An optimization problem that seeks to minimize travel time subject to a set of dynamic and static constraints related to the motion of a vehicle, is formulated. CAVs were simulated using the Rakha-Pasumarthy-Adjerid (RPA) car-following model, which was recently developed. The VT-Micro model is used to estimate fuel consumption and emissions, and the experiments were conducted using the INTEGRATION traffic simulator software. Simulation tests are performed for an intersection and considering various levels of congestion, and included the cases of a roundabout, stop sign and a traffic signal controlled intersection. For the roundabout case, savings in terms of delay and CO2 emissions were obtained up to 80 and 40%, respectively. Findings are promising, however, more efficient algorithms should be developed to account for real-time information.

In [26], a multi-vehicle coordination heuristic approach is developed to optimize the traffic flow of CAVs at traffic signals such that it minimizes system emissions. Impacts of the proposed control strategies are evaluated through simulation experiments in terms of emissions, travel time, and wait time. Scenarios for different penetration levels of the V2X technology and CAVs are explored, considering increasing traffic volume levels. Simulation was conducted through a SUMO traffic simulator on a road with two traffic signals, and emissions were estimated also using the SUMO model of the Handbook Emission Factor for Road Transport (HBEFA). A first observation is that lower penetration rates of V2X technology vehicle control algorithms increase CO2 emissions compared to the baseline, for all volume scenarios. Findings suggested higher volume provide higher emission savings through the coordination heuristic, representing up to 6–13% fewer emissions for lower and higher traffic volumes, respectively.

In [7], it is proposed that a dynamic multi-objective eco-routing strategy for CAVs is implemented in a distributed traffic management system. The dynamic distributed multi-objective eco-routing system takes into account both travel time and environmental pollution impacts. All scenarios assumed 100 %CAVs on the network except for the baseline. The aim was to evaluate the effects of incorporating travel time, idling penalty, and GHG emission in the routing objective function. CAVs were simulated using the well-known Intelligent Driver Car-Following Model (IDM) in which car-following parameters regarding reaction time and minimum safe distance are set to be half of the CVs since CAVs should perceive the changes in the surrounding environment quicker. Some parameters, such as maximum acceleration and deceleration, depend on the engine technology and vehicle weight, thus, the authors calibrate both CAVs and CVs to have the same values for these parameters. The routing of CAVs is mostly focused on communication and information intersection by intersection, where CAVs send and receive information on the destination and best route. Under the agent-based microsimulation platform, network vehicle types are simulated based on the on-road vehicle fleet composition. The MOtor Vehicle Emission Simulator (MOVES) was used to emulate emission sensors and estimate an emission rate based on instantaneous speed and acceleration. The monetary value of GHG emissions is converted into minutes. The suggested routing strategy minimizes both travel time and pollutant emissions. Results show that considering all different scenarios, significant savings in terms of system travel time and system GHG and NOx emissions are in the order of 29–41%, 32–43%, and 12–19%, respectively. Maximum and average concentrations for NOx are also estimated (using the atmospheric air pollution dispersion model SIRANE) and show that the proposed routing strategy can yield significant improvements in air quality, in particular, for the idling penalty-integrated route. Nevertheless, the evaluated scenarios neither consider the transition phase where CVs share the roads with CAVs nor the associated impacts derived from road grade variations.

In [45], a two-level optimization model is developed to assess the impact on gasoline consumption and CO2 emissions of mixed traffic flow with CAVs at an isolated intersection. At the upper level, a dynamic program (DP) is considered to optimize traffic signal timing based on predicted vehicle arrival, while the lower level problem involves optimizing vehicle trajectories such that the gasoline consumption is minimized, which is done based on a model predictive control (MPC) method. The considered case study involved a typical four-arm intersection that is further coded into a simulation platform. Three scenarios regarding different volume levels are explored and for each one different market penetration rates for CAVs are considered, yielding a total of 15 scenarios. The simulation results showed CAVs the gradual introduction can reduce CO2 emissions significantly, and the higher the traffic demand, the lower the performance of the CAV-based control method. For a full CAVs scenario, the reduction in vehicle delay, gasoline consumption, and CO2 emissions can be up to 57%, 22%, and 19%, respectively.

In [43], an online bi-level eco-driving control strategy for connected and automated hybrid electric vehicles (CAHEVs) is developed for both highway and urban environments. The main goal of the model predictive control (MPC)-based strategy is to explore the eco-driving potential of CAHEVs to improve safety, energy consumption and reduce emission, yielding a multi-objective optimization problem. A hybrid platform integrating MATLAB® and SUMO simulator is used in a case study involving a typical medium-sized city. The IDM is used for modeling the CAHEVs. Simulation results show that strategy can reduce the fuel consumption by 34% and also the HC, CO, and NOx emissions up to 25%, 72%, and 30%, respectively.

A cooperative eco-driving system is proposed in [44] focused on reducing energy consumption and pollutant emissions along a corridor with signalized intersections. The proposed system based on new role transition and longitudinal control models is capable of modeling the interactions between different vehicle types. CAVs are simulated using the IDM and emissions are estimated using MOVES. The simulation is conducted under the PTV VISSIM microscopic simulation software and considers different scenarios of CAVs market penetration rates. Results demonstrate that in the 100% CAVs scenario, it can be expected savings in terms of energy consumption around 7% and up to 59% on pollutant emissions, namely, in terms of CO, CO2, HC, and NOx.

A Personal Rapid Transit (PRT) system scheme is proposed in [11] based on green supply chain management (GSCM). The goal is to develop management strategies capable of improving the performance of the PRT transport network, and minimizing system CO2 emissions. The autonomous car fleet is assumed to be electric, which clearly poses a decrease in the operational-related emissions. The scheme involves the central control system, PRT stations and dispatch rules. In the proposed vehicle routing problem, the cost function of the total carbon emissions takes into account energy consumption under three actions: empty movement, embarking/disembarking passengers, and movement with passengers. The Analogic java-based simulation software was used to test the proposed scheme where both the behavior of vehicles and passengers in the PRT system were considered on a network involving 200 fleet vehicles. The environmental impacts of two waiting and reactive strategies were evaluated with the developed PRT system. Results suggested the PRT system which is applied with the first strategy provides emission savings of around 8% on the empty movements.

In [16], an optimal controller is proposed based on an eco-approach system for an isolated signalized intersection that seeks to optimize traffic flow by controlling the speed in a partial CAVs environment. It involves two different modules that are activated either the vehicle is a CV or a CAV. In essence, they propose a conventional vehicle trajectory prediction module, based on the IDM car-following model which is adapted to predict the acceleration profile of CVs, and an optimal controller for CAVs module, which optimizes the CAVs speed profile. In this latter module, system fuel consumption and emissions are minimized, while maintaining the throughput at its optimum level. The VT-Micro fuel consumption and emissions model developed by Rahka et al. (2004) [31] is used in this study. An iterative Pontryagin Minimum Principle procedure is proposed to solve the optimization problem. An integrated VISSIM-MATLAB and

Excel VBA micro-simulation platform were developed to test the proposed controller in a hypothetical simple signalized intersection. Scenarios that took into account different levels of congestion and CAV market penetration rate were defined. Compared to the baseline scenario, the proposed system provides fuel consumption benefits that grow with the increase of congestion level and emissions reductions that grow with MPR and congestion level. Findings revealed that the controller presents up to 58% and 33% savings in fuel and emissions, respectively, and improves by up to 11%.

Later, following this study, the authors in [15] developed an eco-approach and departure controller system to optimize CAVs trajectories at fixed-time signalized intersections to improve fuel efficiency. A control structure involving three main modules was proposed: speed trajectory optimization based on Pontryagin Minimum Principle; Feasible speed trajectory query; and Conventional vehicle trajectory Prediction. A simulation experiment was conducted for a hypothetical signalized intersection to test the proposed trajectory optimization system. Different scenarios involving two levels of congestion and a range of market penetration rates of CAVs were defined. In [16], fuel consumption and CO2 emission are estimated through the VT-Micro model. Simulation results suggest CO2 savings can be achieved for any CAVs scenario, being more evident with the increase of the congestion level and market penetration rates of CAVs ranging between 12 and 16% for the lowest and highest levels of congestion, respectively. It can be observed that for CAVs higher penetration rates, the CO2 emissions become stable.

A two-layer traffic control is proposed in [39] to optimize the traffic flow at a junction aiming to reduce emissions in the junction area and minimize queues in all road links. The focus here is to remove traffic lights at the intersection and explore the potential of CAVs to act cooperatively. This is achieved by developing a nonlinear model predictive control scheme. A network of four junctions and 12 bidirectional roads was chosen as a case study. To test the capability of the proposed control strategy, a microscopic traffic simulation within SUMO and MATLAB is developed. Emissions are estimated in SUMO by COPERT IV emission model at the macroscopic scale. The default microscopic (Krauss) car-following model in SUMO was adopted for simulating CAVs. Simulation results showed that the two-level optimization controller allowed an improvement in the traffic flow of more than 35% and CO2 emission savings around 25% compared to the baseline solution.

In [37], a Green Light Optimal Speed Advisory (GLOSA) system is proposed, and its impacts on the traffic flow and CO2 emissions are analyzed. It is assumed that a connection exists between vehicles and an upcoming traffic light and a simulation platform is developed. Three scenarios were defined based on assistance time: full assistance independently of the signal on, partial assistance when a signal is amber or red, and given full assistance with a lower speed limit. Simulation-based results showed a reduction of up to 10% in CO2 emissions and fuel consumption can be achieved in the case of the third scenario, even for higher demand.

In [9], the authors explored the impacts of CAVs introduction on different networks with traffic signal intersections. Two hypothetical networks of varying complexity were investigated considering two traffic signal control strategies (fixed and adaptive), under varying market penetration rates of CAVs. Simulations were conducted using the SUMO traffic simulator, where the updated version of the Krauss car-following model was adopted for simulating CAVs driving behavior, while an imperfection parameter was included to simulate CVs. For estimating the pollutant emissions, the HBEFA3 based emission model within SUMO was selected. Results on the first scenario, with fixed time signalized intersections, showed that for the case of the simpler network, the market penetration rate of CAVs of 75% yielded lower emission savings, when compared to other penetration rates, that in general provided emission savings ranging between 25–28%. Regarding the second hypothetical network of higher complexity in terms of intersection geometry, a gradual increase in emission savings as the penetration increases (up to 17%) was revealed. Concerning the second scenario of adaptive signalization, simulation results showed that for the full CAVs case, savings of over 22% can be achieved.

The authors from [46] develop an integrated VSL control strategy for CAVs based on the cell transmission model (CTM). The optimal VSL control problem is solved applying genetic algorithms. The control strategy is tested on a freeway corridor with various bottlenecks and scenarios defined for different levels of control strategies, up to a market penetration rate of 10% of CAVs. AVs and CAVs car-following driving behaviors were simulated using IDM (parameters were selected on the basis of existing studies), and lane-changing using the default VISSIM model. An integrated VISSIM-MATLAB platform was developed to evaluate the network impacts. Trucks were also considered in the simulation taking into account local fleet composition. Results show savings up to 7.8%, 9.1% and 5.9% can be achieved for CO2, NOx and PM, respectively.

In [35], the impacts of a small penetration rate (5%) of AVs on the traffic flow and emissions are evaluated by focusing on the effect of stop-and-go traffic waves. Three dampening traffic waves control strategies with the goal to stabilize system traffic flow are developed. Four scenarios of different vehicle fleets are considered (no electric vehicles; residual number; 2030 projection of 3.4%, and 2050 projection of 80% of electric vehicles). This study uses experimental data of vehicle trajectories of the Cognitive and Autonomous Test (CAT) Vehicle (University of Arizona), a highly instrumented and actuated passenger car. Emissions are estimated by applying the MOVES model. Results showed that the most effective controller lies on commanding the AV to select a safe speed and move smoothly, providing higher emissions savings considering the first scenario, which surprisingly yields very similar results on the projected scenarios for 2030 and 2050 (average reductions for CO2, HC, CO and NOx around 30–31%,

50.4–51.6%, 39.1–39.3% and 72.5–73.6%, respectively). One limitation is that the study just focused on one lane of traffic, which is not so realistic for instance, highway sections.

Floating car data-based adaptive traffic signals benefits in terms of energy and air quality impacts at an intersection were explored in [2]. The authors considered a cooperation-competition approach. A greedy algorithm was proposed to optimize the traffic light settings, based on the evaluation of the number of connected vehicles. Experiments were performed in three different scenarios with varied penetration rates of connected vehicles. The VISSIM Wiedemann 99 car-following model was adopted to simulate driving behavior of CAVs. Emissions are estimated using the regression-based model developed by Panis that takes into account instantaneous speed and acceleration of vehicles. The experiments were performed for a typical intersection of four approaching roads. From the results obtained through vehicle trajectories derived from microsimulation, it was found that an increase in the percentage of connected vehicles provided an improvement in traffic signal setting that resulted in a reduction up to 24%, 32%, 21% in average CO2, NOx, and PM emissions, respectively.

In [13], a control framework is proposed to deal with freeway bottlenecks. A model predictive control integrating variable speed limits (VSL) and lane-changing (LC) are developed and a Genetic Algorithm-based strategy is proposed to solve it. Experiments were conducted for a case study of a two-lane freeway, using the open-source microscopic traffic simulation platform SUMO. Vehicles were simulated using the Krauss car-following model. Different traffic demands and incident scenarios were explored. Average emissions rates were estimated based on the HBEFA3 model. Simulation results show that the proposed control method can improve road capacity, fuel consumption and average CO2, PM, and NOx emissions up to 48%. However, the impact of different CAVs penetration rates were not investigated.

The authors from [17] propose a general model framework of mixed traffic under a cyber physical perspective. For that purpose, a stability condition is derived based on linear stability analysis. The main aim is to explore the impact of different CAV penetration rates, number of preceding vehicles on mixed traffic stability. CAVs are simulated using the IDM. Since emissions are correlated with fuel consumption, the latter is estimated following a model for passenger vehicles and considering a third-order polynomial equation mainly based on speed and acceleration. Then, emissions are estimated through the proportional relationship with fuel consumption. Simulation results on a single-lane ring road show that the proposed model can better reflect the real advantages of CAV in mixed traffic when compared to another model. In particular, it is shown that CAVs can improve the stability of mixed traffic, traffic efficiency and fuel consumption, and CO2 emissions can be reduced to half. Best results are obtained for full CAVs environment, that is the higher the CAV market penetration rate, the higher the fuel efficiency.

In [14], an improved car-following model with multiple time lags is proposed to account for different lag patterns on different vehicle types. Based on the Lyapunov function theory, a sufficient condition of the traffic flow stability is established and a controller for mixed traffic is designed and solved through a linear matrix inequality approach. Different market penetration rates of autonomous vehicles (AVs) are defined, and impacts are evaluated. No details are presented regarding the simulation environment or nature of the case study. The reported simulation results revealed the capability of significantly reducing fuel consumption (up to 3%) and CO, HC, and NOx emissions (up to 6%), with minimal values for the 100% scenario of AVs. However, the authors just focused on single-lane roads and very simple driving behaviors/situations that might be not so realistic.

2.3.2 Model adjustment framework

Mattas et al. (2018) [25] evaluated the traffic flow and pollutant emission impacts of introducing AVs and CAVs on a road network through microsimulation. The objective was to reveal the possible benefits of connectivity. The simulation experiment was conducted using the AIMSUN traffic simulation model for a ring road. Three types of vehicles were modeled: the default AIMSUM modified Gipps' car-following model was adopted for reflecting CVs driving behavior, for modeling AVs, the model proposed by [34] was used and finally, for CAVs the model suggested in [38] was applied. CO2 and NOx emissions are estimated using the COPERT model. A total of 21 scenarios of varied market penetrations of each vehicle type were considered for three different traffic demands. Findings indicated that high CAVs penetration rates provided lower emissions in all cases. In particular, for the scenario of higher traffic volume, the opposite occurred for AVs, in which as the market penetration rate increase, it deteriorates the average of CO2 and NOx emissions, 163–181 g/km and 0.594–0.657 g/km, respectively.

In [36], the impacts on GHG emissions and traffic flow performance of different AV driving behaviors are investigated. Different scenarios are defined under high and low traffic demand and considering market penetration of AVs and electrification. Experiments were conducted for highway and arterial roads and the VISSIM microscopic traffic simulation tool was chosen. The default VISSIM parameters were used to represent CVs, while the Wiedemann 99 car-following parameters were subjected to tuning to represent AVs driving behavior. Emissions were estimated through the MOVES model. Simulation results indicate that aggressive AVs can reduce emissions up to 26% on the highway, while cautious AVs can deteriorate traffic performance increase emissions up to 35%. The results show that there is no general impact, and local conditions should be taken into consideration to better support policymaking.

In [4], the authors focused on evaluating the traffic performance and CO2 system-wide emissions of AVs introduction based on an integrated traffic microsimulation and emission modeling approach. To understand more clearly the effect at the operational level, it is assumed that AVs present the same size of engine and fuel technologies as CVs. SUMO microsimulation model to conduct the network experiments and the microscopic Newton-based GHG model (NGM) to estimate emissions are used. Simulation results indicate that highest AVs penetration rates provide better traffic performance, and there were cases where low market penetration rates of AVs yielded worse outcomes than the baseline scenario. In this latter case, the total CO2 emissions increase up to 2.5% when compared to the baseline. Emission reductions are greater for peak-hours. The highest savings in system-wide emissions (4.5%) are achieved for the full AV scenario.

Correia et al. (2019) [5] investigated the effect of introducing 30% of CAVs in an urban fleet on air quality. The study focused on the main avenue of a medium-sized city, surrounded by residential buildings and schools areas, and on evaluating the impact on NOx emissions for a typical weekday. A modeling platform integrating traffic, emission and concentration modeling and estimation as proposed. CAVs were simulated using the VISSIM Wiedemann 99 car-following model with parameters adjusted based on [36]. Vehicle-specific Power (VSP) methodology was applied for different types of vehicles to estimate pollutant emissions and the Computational Fluid Dynamics (CFD) model VADIS (pollutant DISpersion in the atmosphere under VAriable wind conditions) was applied to evaluate air quality in terms of emission concentrations. Two CAVs scenarios were considered: one assuming current engine technology also for CAVs fleet, and another considering that CAVs are battery electric vehicles. On the one hand, simulation results indicated a slight increase in NOx emissions for both CVs and CAVs in the first scenario, also deteriorating air quality with an increase in NOx concentrations. On the other hand, when CAVs are turned into electric, a significant improvement in air quality can be achieved with NOx concentration reduction at morning peak-hour up to 8%.

In [24], a microsimulation study on the impacts of automation and connectivity has on traffic flow and emissions on a highway network is presented. In particular, different vehicle technologies are simulated through are different driver models. CVs were simulated using the Gipps Model, while a combination of the model proposed by [34] and the default AIMSUN model was considered for AVs. Regarding CAVs simulation, it is suggested a combination of the model of longitudinal movement proposed in [38] and the AIMSUN lane changing default model. Emissions are estimated using the well-known EMEP/EEA methodology and the CO2MPAS generic model. Four scenarios were defined based on the full penetration rate of each vehicle type, where one of them presents an increased of 20% in traffic demand. The other four scenarios were defined using the same models for car-following but a vehicle dynamics-based model for the free-flow terms. The simulation was conducted in the AIMSUN traffic simulation software. Results for the fixed traffic demand scenarios revealed that AVs yield the highest emission values per kilometer, while CAVs improve capacity, but, generate more emissions, especially, at peak-hour. Overall results show the differences that in emissions per kilometer in all the basic scenarios do not exceed 6%.

The traffic performance and environmental impacts of AVs on an urban freeway corridor in a metropolitan area were evaluated in [40]. The study is mainly focused on a microscopic simulation modeling of the vehicle's interaction in a mixed context, where the used AVs car-following parameters are based on those reported in [36]. The Wiedemann 99 model was used through VISSIM to simulate both CVs and AVs driving behaviors. For estimating the pollutant emissions, the VSP methodology was used and representative vehicles of the local fleet were considered. Four market penetration rates of AVs were considered. The simulation was conducted for a typical afternoon peak-hour. Results demonstrated that a reduction in emissions up to 5% can be expected at corridor-level with AVs introduction, while travel time increases around 13% for both CVs and AVs. Although there are some limitations in the study, for instance, regarding real on-road emission measurements for current traffic conditions, the findings can be of particular interest for policymakers in the sense of anticipating impacts of AVs operation on CVs behavior.

A recent study exploring the air quality impact of autonomous vehicles on the main avenue of a medium-sized city was presented in [32]. The research focused on NOx and CO2 and it was conducted on a baseline scenario (full of CVs), and two other scenarios considering a 30% market penetration rate of internal combustion engines and electric AVs. The microscopic traffic model PTV VISSIM was used to simulate the introduction of AVs in the traffic network. The widely used Wiedemann 99 Model was chosen to simulate AVs, in which car-following parameters were selected and tuned to represent the potential of AV operations based on existing literature (e.g., [36]). Pollutant emissions were estimated using the VSP methodology, which is capable of reflecting microscopic changes in driving behavior. Regarding the air quality modeling, the CFD model VADIS was applied. Simulation results showed that there can be an increase of around 2% in NOx and 1% in CO2 atmospheric emissions, which can be justified by the influence. AVs had on CVs, allowing also for an increase in the traffic demand, and resulting in more aggressive acceleration situations. Reductions of 30% in emissions can be achieved for the electric AVs scenario.

Recently, a very complete study concerning the impacts on fuel economy, capacity, queue length, vehicle delay, and vehicle emissions considering varying levels of L4-L5 AVs was proposed in [18]. For this purpose, the VISSIM traffic simulator was used, CAVs were modeled based on the All-Knowing Co-EXist Autonomous Vehicle model can be found in VISSIM, while for CVs, the VISSIM Wiedemann 74 model was adopted. Four different traffic networks were explored, which differ in characteristics, mainly in terms of traffic condition, volumes, singularities. Findings indicated that in terms of fuel consumption, savings were reported at each penetration level for each route. Highest savings (up to

22%) were found for higher levels of AVs on roads and for routes involving singularities. In terms of capacity benefits, these were smaller (2%) in urban roads with the increase in L4-L5 AV penetration rate, when compared to freeway routes (up to 18–25%). Significant savings in average queue length were achieved for all studied routes in all scenarios, yielding higher benefits as the market penetration rate of AVs also increases. Concerning average vehicle delay, simulation results suggested higher savings as the number of AVs grow (5–40%), however, it was found the opposite for one of the routes that it increased up to 60%, due to its geometry and traffic conditions, where the descending trend on average delay only improved with at least 30% of AVs. With respect to pollutant emissions, namely, regarding CO, NOx, and VOC, results showed a general trend on higher emission savings up to 10% as the level of AVs in roads increases. However, for the route that involves few traffic lights, 30% less total emissions were reported for the case of AV penetration rates of 65% and 80%.

There are also some studies exploring the impacts autonomous taxis pose to a city network. For instance, in [23], an agent-based model is developed with the goal of optimizing an autonomous taxi fleet so that system distance traveled is minimized. Nevertheless, environmental impacts regarding energy consumption, GHG and SO2 emissions are analyzed. Scenarios varying fleet size, vehicle types and operation strategies were defined and tested, including shared autonomous vehicles (SAV). The model was implemented in the GAMA platform that was integrated with traffic simulation. Results suggest no significant environmental improvements were achieved. In particular, for the optimized SAV scenario, the system energy consumption, GHG and SO2 emissions were 16, 25, and 10% higher, respectively, than current scenario. Regarding the electric autonomous taxis scenario, the SO2 emissions increased by 560% compared to baseline, although there were slight reductions in GHG emissions.

Literature on the environmental impact of CAV technology in buses is scarce. The study presented in [47] focuses on evaluating the emission and energy impacts of connected and autonomous buses. The proposed VISSIM simulation-based framework uses the Wiedemann 74 model and the MICroscopic model for the Simulation of Intelligent Cruise control (MIXIC) model for simulating CVs and connected and automated buses. CVs model parameters were calibrated by using field survey traffic data. For emission estimation, the VSP methodology is used for both passenger vehicles and buses. The study was conducted in a mixed traffic environment and on an urban expressway with eight lanes. Three main scenarios were considered: the baseline, the dedicated managed lane (only automated buses in a dedicated lane), dedicated lane (all buses—automated or not—in a dedicated lane). Testing different market penetration rates of CAVs are considered under each scenario. The results suggest that emission and energy consumption savings for shares of CAV above 30% were more evident. Regarding the managed lane scenarios, both present benefits. Experiments on increasing buses volume are not explored.

Researchers have also been studying cyber issues associated with connectivity. For instance, in [8], cyber-attack on the traffic flow stability is analyzed through position and speed, and environmental impacts were evaluated. When a CAV (cooperative adaptive cruise control (CACC) system) is under cyber-attack, it will share imprecise data with others. Different cyber-attacked scenarios are explored and evaluated considering indicators for capacity, safety (rear-end collision), emissions (HC, NOx, CO2, CO), and fuel consumption. Simulation experiments on an one-lane road revealed that a cyber-attack area deteriorates safety condition and the increase of cyber-attacked vehicles causes up to 20% more emissions and fuel consumption. These results raise concerns of cyber security.

A topic that also deserves attention is the life-cycle environmental performance of AVs. In [30], the authors focused on a simulation-based approach for evaluating the life-cycle environmental performance assuming a full adoption of battery electric CAVs at the urban mobility level and 15-year vehicle service life. Traffic simulation was conducted for a real road network in the city of Rome, Italy, by using the travel demand modeling software EMME by INRO. CAVs were simulated by assuming a capacity gain of 40% for intra-urban roads and 80% for highways (based on previous literature). Results indicate savings around 33% in travel time, improvements in system speed of 50% and regarding distance traveled, small increase can be achieved for highways (8%) and a decrease for urban roads (5%). Findings suggest that on an average, CAVs present significant reduction (60%) on the environmental impact at the mobility system level, but the highest GHG emissions related to construction, maintenance and end-of-life processes, resulting in 35% higher emissions than the internal combustion engine vehicles, 22% higher than hybrid electric vehicles and a 5% higher than electric vehicles. These findings show that the overall environmental impact should be further explored and improved.

The present survey showed evaluation of environmental impact in terms of noise is relatively scarce in the literature. Pantella et al. (2019) [29] addressed this issue by presenting a simulation-based study within a real city network assuming CAVs noise emissions are modeled as battery electric vehicles noise emissions, i.e., only considering the rolling noise component. The noise was estimated by using the Common Noise Assessment Methods in Europe (CNOSSOS-EU) which involves two noise components: rolling noise (due to the tyre/road interaction) and propulsion noise (due to engine technology). Simulation is performed using the framework already described above (such as in [30], in particular, 100% CAVs scenario was simulated assuming a capacity gain of 40% for intra-urban roads and 80% for highways. Results suggested that a reduction of almost 25% and 6% in noise emissions on urban roads and highway sections can be achieved, respectively, showing a beneficial effect on the noise pollution. However, a closer link analysis revealed some negative impacts on noise emissions due to an increase in traffic volume and speed.

2.4 Conclusions and Future Research Directions

Although few pilot cases have been running in some cities around the world, CAVs deployment on roads are expected to be disruptive. Such new advanced technologies bring both opportunities and challenges. Considerable amount of research has been conducted in assessing the traffic-related impacts of such technologies. However, literature review shows safety and traffic efficiency improvements as the main drivers for automation development. But anticipating CAVs environmental-related impacts are fundamental for sustainable and smart cities. In fact, it is important to rethink in a more holistic way about the potential environmental changes the deployment of new emerging technologies pose to the road traffic networks.

There is no doubt that OR has and will bring important contributions to improve operations and the environmental impacts of road traffic. Particularly in the case of CAVs introduction, research has been mainly focused on applying simulation, (multi-objective) optimization, and control strategies. Based on the literature, we can conclude that establishing scenarios under simulation and exploring vehicle interactions and consequent driving behaviors are an insightful lens for examining the impacts of CAVs in a mixed traffic network. These can be used to support multi-criteria decision making and aid traffic engineers and policymakers to better define strategies to efficiently accommodate CAVs in the future. In general, the environmental benefits of CAVs sharing the roads with CVs become more significant as the market penetration rate increases. However, some results point to marginal or negative environmental benefits with the introduction of CAVs, even at high market penetration rates (see, e.g., [4]). Nevertheless, surveyed results show that introducing CAVs can potentially reduce pollutant emissions, but trade-offs can be achieved, mostly between emissions reduction and travel time. Studies suggest that in a mixed context, CAVs can influence near-by CVs driving behavior, controlling them indirectly. This is a very important finding especially for the transition period, since an CAV-eco-routing strategy can induce a similar eco-driving behavior on CVs. These findings can be useful for policymakers that design strategies to reduce emissions depending on specific traffic conditions.

Given the plethora of specificities and traffic and road conditions, more research efforts are still needed. How to optimize CAVs driving behavior while approaching an intersection has been the subject of much research, but either considering improvement of multiple components, or impacts related to CVs and CAVs interactions, are not yet fully exploited [3, 4], and thus, these should be well studied and understood before their deployment. Based on the relevant literature sample we can say that evaluation of environmental impacts at system level is relatively scarce. Future work should also take into consideration how increasing market penetration rates of electric vehicles and other alternative fuels, and considering different vehicle types, will impact traffic, comfort, noise, and pollutant emissions and consequently, air quality in an integrated way. Most studies assume a constant travel demand, which can be not so realistic, considering the fact that CAVs platooning and driving behaviors may increase capacity, as well as possibly increase of traffic volume due to induced demand related to, e.g., underserved populations, can pose variation on travel demand, which will certainly have consequent environmental impacts.

Acknowledgements

The authors acknowledge the support of projects: UIDB/00481/2020 and UIDP/00481/2020-FCT; CENTRO-01-0145-FEDER-022083; InFLOWence (POCI-01-0145-FEDER-029679), co-funded by COMPETE2020, Portugal2020 - Operational Program for Competitiveness and Internationalization (POCI), European Union's ERDF (European Regional Development Fund) and the Portuguese Foundation for Science and Technology (FCT - Fundação para a Ciência e a Tecnologia); and Interreg Europe Project PriMaaS PGI05830.

References

[1] R. Arvin, A.J. Khattak, M. Kamrani and J. Rio-Torres. Safety evaluation of connected and automated vehicles in mixed traffic with conventional vehicles at intersections. *Journal of Intelligent Transportation Systems*, 25(2): 170–187, 2021.

[2] V. Astarita, V.P. Giofrè, G. Guido and A. Vitale. A single intersection cooperative-competitive paradigm in real time traffic signal settings based on floating car data. *Energies*, 12(3): 409, 2019.

[3] Y. Bichiou and H.A. Rakha. Developing an optimal intersection control system for automated connected vehicles. *IEEE Transactions on Intelligent Transportation Systems*, 20(5): 1908–1916, 2019.

[4] J. Conlon and J. Lin. Greenhouse gas emission impact of autonomous vehicle introduction in an urban network. *Transportation Research Record*, 2673(5): 142–152, 2019.

[5] L.P. Correia, S. Rafael, D. Lopes, J. Bandeira, M.C. Coelho, M. Andrade, C. Borrego and A.I. Miranda. Assessment of local air quality for different penetration levels of connected autonomous vehicles. 186: 153–162, 2019.

[6] L. Cui, H. Jiang, B.B. Park, Y.-J. Byon and J. Hu. Impact of automated vehicle eco-approach on human-driven vehicles. *IEEE Access*, 6: 62128–62135, 2018.

[7] S. Djavadian, R. Tu, B. Farooq and M. Hatzopoulou. Multi-objective eco-routing for dynamic control of connected and automated vehicles. *Transportation Research Part D: Transport and Environment*, 87: 102513, 2020.

[8] C. Dong, H. Wang, D. Ni, Y. Liu and Q. Chen. Impact evaluation of cyber-attacks on traffic flow of connected and automated vehicles. *IEEE Access*, 8: 86824–86835, 2020.

[9] İ.G. Erdağı, M.A. Silgu and H.B. Çelikoğlu. Emission effects of cooperative adaptive cruise control: A simulation case using sumo. 62: 92–100, 2019.

[10] D.J. Fagnant and K. Kockelman. Preparing a nation for autonomous vehicles: Opportunities, barriers and policy recommendations. *Transportation Research Part A: Policy and Practice*, 77: 167–181, 2015.

[11] E. Fatnassi and J. Chaouachi. Dynamic carbon emissions minimization for autonomous vehicles in the context of on-demand transportation systems. 2016(August): 698–703, 2016.

[12] J.M. Greenwald and A. Kornhauser. It's up to us: Policies to improve climate outcomes from automated vehicles. *Energy Policy*, 127: 445–451, 2019.

[13] Y. Guo, H. Xu, Y. Zhang and D. Yao. Integrated variable speed limits and lane-changing control for freeway lane-drop bottlenecks. *IEEE Access*, 8: 54710–54721, 2020.

[14] L. Huang, C. Zhai, H. Wang, R. Zhang, Z. Qiu and J. Wu. Cooperative adaptive cruise control and exhaust emission evaluation under heterogeneous connected vehicle network environment in urban city. *Journal of Environmental Management*, 256: 109975, 2020.

[15] H. Jiang, S. An, J. Wang and J. Cui. Eco-approach and departure system for left-turn vehicles at a fixed-time signalized intersection. *Sustainability (Switzerland)*, 10(1): 273, 2018.

[16] H. Jiang, J. Hu, S. An, M. Wang and B.B. Park. Eco approaching at an isolated signalized intersection under partially connected and automated vehicles environment. *Transportation Research Part C: Emerging Technologies*, 79: 290–307, 2017.

[17] S. Jin, D.-H. Sun, M. Zhao, Y. Li and J. Chen. Modeling and stability analysis of mixed traffic with conventional and connected automated vehicles from cyber physical perspective. *Physica A: Statistical Mechanics and its Applications*, 551: 124217, 2020.

[18] O. Kavas-Torris, M.R. Cantas, K. Meneses Cime, B. Aksun Guvenc and L. Guvenc. The effects of varying penetration rates of l4-l5 autonomous vehicles on fuel efficiency and mobility of traffic networks. *SAE Technical Papers*, 2020-April 2020.

[19] F. Leach, G. Kalghatgi, R. Stone and P. Miles. The scope for improving the efficiency and environmental impact of internal combustion engines. *Transportation Engineering*, 1: 100005, 2020.

[20] S. Lee, E. Jeong, M. Oh and C. Oh. Driving aggressiveness management policy to enhance the performance of mixed traffic conditions in automated driving environments. *Transportation Research Part A: Policy and Practice*, 121: 136–146, 2019.

[21] C. Legacy, D. Ashmore, J. Scheurer, J. Stone and C. Curtis. Planning the driverless city. *Transport Reviews*, 39(1): 84–102, 2019.

[22] G. Lozzi, M. Rodrigues, E. Marcucci, V. Gatta, T. Teoh, C. Ramos and E. Jonkers. Research for TRAN committee—sustainable and smart urban transport. *European Parliament, Policy Department for Structural and Cohesion Policies, Brussels*, 2020.

[23] M. Lu, M. Taiebat, M. Xu and S.-C. Hsu. Multiagent spatial simulation of autonomous taxis for urban commute: Travel economics and environmental impacts. *Journal of Urban Planning and Development*, 144(4): 04018033, 2018.

[24] M. Makridis, K. Mattas, C. Mogno, B. Ciuffo and G. Fontaras. The impact of automation and connectivity on traffic flow and CO2 emissions. A detailed microsimulation study. *Atmospheric Environment*, 226: 117399, 2020.

[25] K. Mattas, M. Makridis, P. Hallac, M.A. Raposo, C. Thiel, T. Toledo and B. Ciuffo. Simulating deployment of connectivity and automation on the Antwerp ring road. *IET Intelligent Transport Systems*, 12(9): 1036–1044, 2018.

[26] K. McConky and V. Rungta. Don't pass the automated vehicles!: System level impacts of multi-vehicle CAV control strategies. *Transportation Research Part C: Emerging Technologies*, 100: 289–305, 2019.

[27] D. Milakis, B. van Arem and B. van Wee. Policy and society related implications of automated driving: A review of literature and directions for future research. *Journal of Intelligent Transportation Systems*, 21(4): 324–348, 2017.

[28] A. Olia, S. Razavi, B. Abdulhai and H. Abdelgawad. Traffic capacity implications of automated vehicles mixed with regular vehicles. *Journal of Intelligent Transportation Systems*, 22(3): 244–262, 2018.

[29] S.M. Patella, F. Aletta and L. Mannini. Assessing the impact of autonomous vehicles on urban noise pollution. *Noise Mapping*, 6(1): 72–82, 2019.

[30] S.M. Patella, F. Scrucca, F. Asdrubali and S. Carrese. Carbon footprint of autonomous vehicles at the urban mobility system level: A traffic simulation-based approach. *Transportation Research Part D: Transport and Environment*, 74: 189–200, 2019.

[31] H. Rakha, M. Snare and F. Dion. Vehicle dynamics model for estimating maximum light-duty vehicle acceleration levels. *Transportation Research Record*, 1883(1): 40–49, 2004.

[32] S. Rafael, L.P. Correia, D. Lopes, J. Bandeira, M.C. Coelho, M. Andrade, C. Borrego and A.I. Miranda. Autonomous vehicles opportunities for cities air quality. *Science of the Total Environment*, 712: 136546, 2020.

[33] SAE On-Road Automated Vehicle Standards Committee. Taxonomy and definitions for terms related to driving automation systems for on-road motor vehicles. SAE International: Warrendale, PA, USA, 2018.

[34] S.E. Shladover, D. Su and X.-Y. Lu. Impacts of cooperative adaptive cruise control on freeway traffic flow. *Transportation Research Record*, 2324(1): 63–70, 2012.

[35] R.E. Stern, Y. Chen, M. Churchill, F. Wu, M.L. Delle Monache, B. Piccoli, B. Seibold, J. Sprinkle and D.B. Work. Quantifying air quality benefits resulting from few autonomous vehicles stabilizing traffic. *Transportation Research Part D: Transport and Environment*, 67: 351–365, 2019.

[36] C. Stogios, D. Kasraian, M.J. Roorda and M. Hatzopoulou. Simulating impacts of automated driving behavior and traffic conditions on vehicle emissions. *Transportation Research Part D: Transport and Environment*, 76: 176–192, 2019.

[37] H. Suzuki and Y. Marumo. A new approach to green light optimal speed advisory (GLOSA) systems for high-density traffic flow. 2018(November): 362–367, 2018.

[38] A. Talebpour and H.S. Mahmassani. Influence of connected and autonomous vehicles on traffic flow stability and throughput. *Transportation Research Part C: Emerging Technologies*, 71: 143–163, 2016.

[39] T. Tettamanti, A. Mohammadi, H. Asadi and I. Varga. A two-level urban traffic control for autonomous vehicles to improve network-wide performance. *Transformation Research Procedia*, 27: 913–920, 2017.

[40] R.F. Tomás, P. Fernandes, E. Macedo, J.M. Bandeira and M.C. Coelho. Assessing the emission impacts of autonomous vehicles on metropolitan freeways. *Transformation Research Procedia*, 47: 617–624, 2020.

[41] N.J. Van Eck and L. Waltman. Software survey: Vosviewer, a computer program for bibliometric mapping. *Scientometrics*, 84(2): 523–538, 2010.

[42] A. Wang, C. Stogios, Y. Gai, J. Vaughan, G. Ozonder, S. Lee, I.D. Posen, E.J. Miller and M. Hatzopoulou. Automated, electric, or both? Investigating the effects of transportation and technology scenarios on metropolitan greenhouse gas emissions. *Sustainable Cities and Society*, 40: 524–533, 2018.

[43] S. Wang and X. Lin. Eco-driving control of connected and automated hybrid vehicles in mixed driving scenarios. *Applied Energy*, 271: 115233, 2020.

[44] Z. Wang, G. Wu and M.J. Barth. Cooperative eco-driving at signalized intersections in a partially connected and automated vehicle environment. *IEEE Transactions on Intelligent Transportation Systems*, 21(5): 2029–2038, 2020.

[45] Z. Yao, B. Zhao, T. Yuan, H. Jiang and Y. Jiang. Reducing gasoline consumption in mixed connected automated vehicles environment: A joint optimization framework for traffic signals and vehicle trajectory. *Journal of Cleaner Production*, 265: 121836, 2020.

[46] M. Yu and W.D. Fan. Optimal variable speed limit control in connected autonomous vehicle environment for relieving freeway congestion. *Journal of Transportation Engineering Part A: Systems*, 145(4): 04019007, 2019.

[47] Y. Zhang, X. Chen and L. Yu. Evaluating the emission and energy impacts of automated buses on urban expressways. *Transportation Research Record*, 2674(12): 515–529, 2020.

Chapter 3

ELECTRE I for Balancing Projects: Case Studies for Selecting Suppliers and Portfolio Investment Schemes

*Laura Lotero** and *Mario Sergio Gomez-Rueda*

3.1 Introduction

Every day, decisions need to be made both in professional and personal activities. However, the complexity of our globalized and intertwined world, and the need to take into account different criteria, scenarios, or preferences of decision-makers, impose the need to pay close attention to the ability which supports decision-making using mathematical methods. This allows us to foresee and understand both the decision-making process and the consequences of the aforementioned decision. When dealing with a problem with multiple alternatives considered under different criteria, methods of decision analysis falling in the category of Multiple Criteria Decision-Making (*MCDM*) or Multiple Criteria Decision Aid (*MCDA*), which also refers to Multiple Criteria Decision Analysis

Universidad Pontificia Bolivariana.
* Corresponding author: laura.loterov@upb.edu.co

or Multi-Attribute Decision Making, should be used. Note that *MCDA* or *MCDM* are interchangeably used in the literature.

There is a vast corpus of *(MCDA)* methods, their use, applications, differences, and comparisons between the different approaches [22, 9, 12, 17, 11, 18]. Even though different authors might establish different categories for comparing MCDA methods, for discrete problems we can distinguish those solved by pairwise comparisons (indifference or preference of one variant over the other) and those based on outranking methods, although they also can be combined. Among the MCDA outranking methods to classify various alternatives for deterministic and exact data, we have the family of ELECTRE approaches [7, 15].

In this chapter, we will provide some of the milestones in the development of ELECTRE methods for MCDA, we will provide a detailed explanation of the ELECTRE I process, and finally, we will show two case studies based on and adapted from real life applications.

3.2 Milestones in the Family of ELECTRE Methods

The ELECTRE method originated in 1965 as a development of the European consulting firm (SEMA) [8]. At that time, a research team from this organization focused on the analysis of specific problems with multiple criteria in the real world concerning decisions made about the development of new activities in companies. To "solve" these types of problems, a general multi-criteria method, called MARSAN (Methode d'Analyse de Recherche et de S'election d'Activit'es Nouvelles), was built. The analysts used a weighted sum-based technique included in the MARSAN method for the selection of the new activities.

Using this method, SEMA engineers noted some serious inconsistencies in the application of the technique. Therefore, they consulted B. Roy and a new method was found later to overcome the limitations of the MARSAN method. This idea began in 1965, and after a while, it was called ELECTRE I. In that same year (July 1965) it was presented for the first time in a conference on the new method of multiple criteria classification (les journees d'etudes sur les methodes de calculation dans les sciences de l'homme) [6]. The first publications on the advancement of the conceptualization of the ELECTRE method were published as a research report in 1966, the notorious Travail Note 49 from the consultancy SEMA. Shortly after its publication, it became known that ELECTRE I had great success when applied to a wide range of fields, but, the method did not become recognized until 1968 when it was published in RIRO (Revue d'Informatique et de Recherche Op'erationnelle) [15]. The abbreviation of ELECTRE means ELimination Et Choix Traduisant la REalite.

In the late sixties, the need for defining an advertising plan in the process of media planning to answer how to establish a proper classification system for periodicals such as magazines and newspapers led to the birth of ELECTRE II: a method to rank the actions of the problem from the best to the worst [14].

However, in a world where perfect knowledge is diffuse, imperfect knowledge could only be taken into account in ELECTRE methods by using probabilistic distributions and the expected utility criterion. Clearly, more work needed to be done, but, the research covering the area was still in its early stages.

For this reason, another way of dealing with uncertainty, imprecision, and poor determination, the threshold approach has become known and a few years later a new method was devised to classify stocks: ELECTRE III [19]. The new ideas introduced by this method were the use of pseudo criteria and fuzzy binary relations of superior classification. Another ELECTRE method, known as ELECTRE IV [20], arose from a new real-world problem, which was related to the Paris metro network in which it was possible to classify the actions without using the importance coefficients of the relative criteria. This is the only ELECTRE method that does not use this type of coefficient. Furthermore, the new method was equipped with an integrated framework of superior relationships.

The methods created up to this point were specially designed to have a basis for decision making when classifying and choosing actions. However, in the late 1970s a new technique was proposed to classify actions into predefined and ordered categories, that is, the trichotomy procedure [16]. This is a decision tree based approach. Some years later, for supporting decision-making in a large banking company that faced a problem about accepting or rejecting loans requested by companies, a specific method was devised and applied in 10 sectors of activity. The most recent classification method is ELECTRE TRI (ELECTRE tree), which is largely inspired by these earlier works [10]. It removed everything specific given its application context. In fact, this new method is simpler and more general.

3.3 Context of Application of the ELECTRE Methods

ELECTRE methods have a long history of successful real-world applications with considerable impact on human decisions [4]. Several areas of application can be identified, such as: agriculture and forest management, energy, environment and water management, finance, military, project selection (call for tenders), transport, medicine, and nanotechnologies. But as a method, they also have their theoretical limitations. Therefore, when applying ELECTRE methods, analysts must first check if their theoretical characteristics respond to the characteristics of the context in which it will be used [6].

In the ELECTRE methods, two alternatives are established to achieve an over classification through the thresholds of agreement and disagreement; the first quantifies the maximum point to achieve a greater number of attributes and detect which of the two alternatives is the preferred one; the second finds the minimum of existence, where there is no attribute for one alternative to be better than another. The purpose of the method is based on finding the *kernel*: the best ranked and indifferent alternatives.

ELECTRE methods are proven to be plausible when tested in decision situations with the following characteristics [5]:

1. The decision maker must include at least three criteria in the model, however, when the model has five or more criteria, these aggregation procedures work more efficiently.
2. When the evaluation of the actions is carried out on an interval scale, these scales are not adequate to analyze and compare their differences. So it is not easy to define a coding that makes any sense concerning the preferential terms of the proportions.
3. It is difficult to add the criteria on a single scale due to the existing heterogeneity with respect to nature at the time of evaluating between criteria.
4. The reward of a criterion with loss that is given by the gain of another criterion is not acceptable for the model, so these cases require some aggregation procedures.
5. The terms of preference are not affected by small differences of evaluation, otherwise, it happens when many differences of the small range are accumulated. At this time, discrimination thresholds are used, which lead to a structure preferably with binary relationships.

The ELECTRE methods continue to evolve and gain acceptance thanks to new application areas, new methodologies, and theoretical developments, as well as user-friendly software implementations [23, 8].

At present, evolutionary algorithms have been used to solve large-scale problems, as well as to mitigate the complexity of some calculations in the ELECTRE methods, mainly, due to some non-linearities in the formulas used in these methods. Noting that not all ELECTRE methods could be applied to all realities, an evolutionary approach was proposed addressing the construction of superior relationships in the context of existing ELECTRE methods and the environment of the area in which it will be applied [3, 1].

3.4 Theoretical and Conceptual Explanation of ELECTRE I

As indicated in the introductory part of this chapter, we will focus and give a mathematical explanation and elaborate the application examples for the ELECTRE I method.

Generally, in a multi-attribute or multi-criteria decision-making process, it is indicated that alternative **a** is preferred to alternative **b** if, given the level of understanding of the decision-maker preferences as well as the quality of the information regarding all the criteria available to evaluate the alternatives against each other, there are enough arguments to affirm that the alternative **a** is at least as good as the alternative **b** and there are no consistent arguments to show otherwise. Some of the fundamental contributions of the ELECTRE I method in MCDA processes are:

- The construction of a relationship of improvement or overqualification.
- The use of the relationship built.

So, it will be affirmed that alternative **a** over classifies or outranks alternative **b** (or that alternative **a** is superior to alternative **b**), when for the criteria considered, "alternative **a** is equal to or greater than **b** in most of the criteria and when in the remaining criteria, the difference in qualification is not too important".

The main objective of the ELECTRE I method is to achieve a nucleus (N) or Kernel subset of alternatives so that any alternative that does not belong to the set N is surpassed at least by another alternative of N. In other words, it is not a set of favorite alternatives, but a set for which you can find the best compromise solution. Therefore, the ELECTRE I method seeks to acquire a partition of the set of alternatives A, which is a finite set that acquires all the possible alternatives considered and divides them into two subsets N and $A \backslash N$ in which it holds that:

- Each alternative of $A \backslash N$ is surpassed by at least one alternative of N.
- The alternatives of N are incomparable to each other.
- $N \cap A \backslash N$ is the empty set.
- $N \cap A \backslash N$ is the set A.

So, the over classification is formed based on two notions: *concordance* and *discordance*.

- CONCORDANCE quantifies the extent to which alternative **a** is more preferred than **b** for a large number of attributes.
- DISCORDANCE quantifies the extent to which there is no attribute for which **b** is much better than **a**.

3.4.1 Decision matrix

The ELECTRE I method takes the initial information, as a matrix arrangement of the form: alternatives (A_n) versus criteria (C_m). This information is collected by the decision maker according to the context of the problem that she wishes to analyze. For each alternative, the observed values (OV) must be known according to the evaluation criteria. On the other hand, each criterion will have the objective of maximizing or minimizing depending on the context of the criterion

Table 3.1: Generic decisional matrix.

	C_1	C_2	...	C_m
A_1	OV(1,1)	OV(1,2)	...	OV(1,m)
A_2	OV(2,1)	OV(2,2)	...	OV(2,m)
...	...	...	...	...
A_n	OV(n,1)	OV(n,2)	...	OV(n,m)

(For instance: The utility is supposed to be always maximized or the negative impacts on the environment will always be minimized).

For each criterion, the weighting or weight (W_m) that will be in comparison with the other criteria must be known. This information is assigned directly by the decision maker. The AHP method [21] is recommended to determine this value in a more structured way. In case the decision maker does not have specific preferences, she/he must assign the same weight to all W_m.

Table 3.2: Vector of weights for the decision criteria.

C_1	C_2	C_3	...	C_m
W_1	W_2	W_3	...	W_m

where,

$$\mathrm{W} = 1 = \sum_{j=1}^{m} W_j. \tag{3.1}$$

Example

In the selection of the best investment project, there are five alternatives, which were evaluated by four criteria: the Net Present Value (NPV—Financial Criteria), Percentage CO2 Emission (CO2—Environmental Criterion), Employees (Labor Criterion—Social) and the risk index (RI—Risk Criterion).

Table 3.3: Example of the decisional matrix.

Alternative	*NPV (Max)*	*CO2 (Min)*	*Employees (Max)*	*R-I (Min)*
Project 1	\$1200	0.30	35	0.08
Project 2	\$2300	0.33	30	0.27
Project 3	\$1570	0.39	20	0.27
Project 4	\$1330	0.26	44	0.08
Project 5	\$1600	0.40	20	0.25

Table 3.4: Example of the vector of weights for the criteria.

NPV	*CO2*	*Employees*	*R-I*
35.0%	25.0%	25.0%	15.0%

3.4.2 Matrix of concordance indices

The objective of the ELECTRE I method is to obtain the Aggregate Dominance Matrix. In order to do so, it is necessary to recognize the over qualification of *Concordance* and *Disagreement*. In Figure 3.1 you can see the development of the entire calculation with the ELECTRE I method.

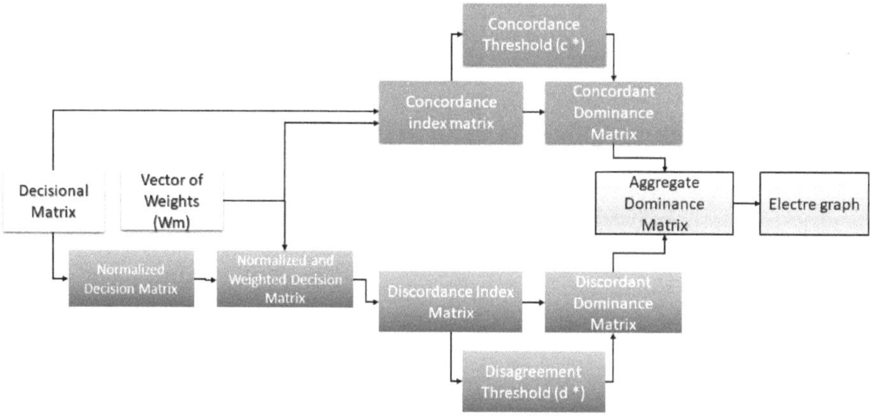

Figure 3.1: Schematic explanation of the calculation process of ELECTRE I.

To perform the calculations of the concordance and disagreement index matrices, the 'OutrankingTools' package set in the free R software can be used. This package was developed by [13]. Alternatively, the J-Electre software developed by [2] can also be used.

In the concordance indices matrix the observed values of the alternatives are compared with each other according to each criterion. This matrix has the following properties:

- It is a square matrix of order *n* (n = number of alternatives).
- The elements of the array are values between 0 and 1.
- The diagonal has no values.

Then, a concordance index is defined for each ordered pair of alternatives (a,b), that is, $C(a,b)$ in the form of

$$C(a,b) = \frac{1}{W} \times \sum_{j/g_j(a) \geq g_j(b)} W_j \tag{3.2}$$

where $g_j(a)$ is the evaluation of the alternative **a** under the g_j criterion.

Table 3.5: Example calculation matrix concordance indices.

	Project 1	*Project 2*	*Project 3*	*Project 4*	*Project 5*
Project 1	-	0.65	0.65	0.15	0.65
Project 2	0.35	-	1	0.35	0.85
Project 3	0.35	0.15	-	0.35	0.50
Project 4	1	0.65	0.65	-	0.65
Project 5	0.35	0.15	0.75	0.35	-

3.4.3 Normalized decision matrix (NMD)

To normalize the decision matrix, we proceed for each element of the matrix, dividing it by the rank of each column (of each criterion). This range is known as the difference from the ideal to the anti-ideal. It is necessary to remember that each criterion makes sense, that is, each criterion will be either seeking to maximize or minimize.

For the example used (see Example in 3.4.1), the normalized decision matrix is:

Table 3.6: Example of normalized decision matrix calculation.

Alternative	*NPV (Max)*	*CO2 (Min)*	*Employees (Max)*	*R-I (Min)*
Project 1	1.091	2.143	1.458	0.421
Project 2	2.091	2.357	1.250	1.421
Project 3	1.427	2.786	0.833	1.421
Project 4	1.209	1.857	1.833	0.421
Project 5	1.455	2.857	0.833	1.316

3.4.4 Normalized and weighted decision matrix (NMDP)

This matrix is obtained by multiplying the weights of each criterion in each Normalized Observed Value of the NMD matrix.

$$\text{NMDP} = \left[NMD_{(n,m)}\right] \cdot \left[W_m\right]. \tag{3.3}$$

For the example used, the normalized and weighted decision matrix is as in Table 3.7.

Table 3.7: Example calculation of the normalized and weighted decisional matrix.

Alternative	*NPV (Max)*	*CO2 (Min)*	*Employees (Max)*	*R-I (Min)*
Project 1	0.382	0.536	0.365	0.063
Project 2	0.732	0.589	0.313	0.213
Project 3	0.500	0.696	0.208	0.213
Project 4	0.423	0.464	0.458	0.063
Project 5	0.509	0.714	0.208	0.197

3.4.5 Matrix of disagreement indices

This matrix defines the discordance between the alternatives so that the quotient is calculated between the greatest difference in absolute value of the criteria for which alternative **a** is alternative **b**, and the greatest difference in absolute value between the results achieved by alternative **a** and **b**. This matrix has the following properties:

- It is a square matrix of order n (n = number of alternatives).
- The elements of the array are values between 0 and 1.
- The diagonal has no values.

So, the discordance index $D(a,b)$ is defined as:

$$D(a,b) = 0 \text{ if } g_j(a) \geq g_j(b), \text{ for all } j = 0,1,2,...,n \tag{3.4}$$

$$D(a,b) = \frac{1}{d} \cdot \{_{(a,b)/(g_j(a)<g_j(b)} max(g_j(a) - g_j(b))\} \tag{3.5}$$

if $g_j(a) < g_j(b)$ for some pair (a,b), where d is the maximum difference for any criteria and any pair of alternatives.

In this way, $D(a,b)$ is an index in which its values are between 0 and 1, and it increases if the preference of alternative **b** over alternative **a** is noticeable for at least one criterion.

For the example provided, the normalized and weighted decision matrix is:

Table 3.8: Example calculation matrix of discordance indices.

	Project 1	*Project 2*	*Project 3*	*Project 4*	*Project 5*
Project 1	-	1	0.3364	0.1182	0.3636
Project 2	0.0045	-	0	0.0127	0
Project 3	0.0136	0.6636	-	0.0218	0.0273
Project 4	0	0.8818	0.2182	-	0.2465
Project 5	0.0136	0.6364	0	0.0218	-

3.4.6 Concordance threshold ($c*$) and discordance threshold ($d*$)

Comparing the concordance and discordance indices, the relationship of overcoming of ELECTRE I will be built. For them, it is necessary to specify their respective limits or thresholds. These thresholds will be levels of agreement and disagreement that reflects the demand and tolerance of the decision maker when formulating the over classification relationship, as it was defined in the initial development of this method.

A good approximation for the calculation of the thresholds can be defined by the mean value of each matrix of indices, that is, $c*$ will be the mean of the matrix of concordance indices and $d*$ will be the mean of the matrix of unconformity indices.

Taking the example that is being done, then, $c* = 0.5275$ and $d* = 0.2289$.

3.4.7 Concordant dominance matrix (CDM) and discordant dominance matrix (DDM)

These matrices are calculated by the exceedance relations S, which can be defined as follows:

$$\text{aSb} \Longleftrightarrow \text{C(a,b)} \geq c*. \tag{3.6}$$

and

$$\text{D(a,b)} \leq d*. \tag{3.7}$$

For those situations where the exceedance ratio is verified, then, the value 1 is assigned.

Taking the values obtained in the calculation example, we could expose these relationships with the following examples, where p_i stands for Project i evaluated:

$C(p_1, p_3)$=0.65 then $C(a,b) \geq 0.5275$. Therefore, the value is 1 in CDM_{p_1,p_3} $C(p_3, p_5)$=0.50. Then, $C(a,b) \geq 0.5275$. Therefore, the value is 0 in CDM_{p_3,p_5}

$D(p_1, p_3)$=0.65. Then, $D(a,b) \leq 0.2289$. Therefore, the value is 0 in DDM_{p_1,p_3} $D(p_3, p_5)$=0.50. Then $D(a,b) \leq 0.2289$. Therefore, the value is 1 in DDM_{p_3,p_5}.

The complete evaluation of the CDM and DDM is presented in Table 3.9 and Table 3.10.

3.4.8 Aggregate dominance matrix (ADM)

In this matrix the concordance and discordance relationships are expressed simultaneously. For them it is indicated that in the equal positions of both matrices or the homologous coordinates, if both values are 1 then said position in the aggregate dominance matrix will take the value of 1, in other cases, then the value will be 0.

Table 3.9: Example calculation of the concordant dominance matrix (CDM).

	Project 1	*Project 2*	*Project 3*	*Project 4*	*Project 5*
Project 1	-	1	1	0	1
Project 2	0	-	1	0	1
Project 3	0	0	-	0	0
Project 4	1	1	1	-	1
Project 5	0	0	1	0	-

Table 3.10: Example calculation of the discordant dominance matrix (DDM).

	Project 1	*Project 2*	*Project 3*	*Project 4*	*Project 5*
Project 1	-	1	0	1	0
Project 2	1	-	1	1	1
Project 3	1	0	-	1	1
Project 4	1	0	1	-	0
Project 5	1	0	1	1	-

It can also be indicated that the mathematical relationship would be:

$$\text{ADM} = [CDM] \cdot [DDM]. \tag{3.8}$$

The calculation of the ADM for the aforementioned example is given by Table 3.11.

Table 3.11: Example calculation of the aggregate dominance matrix (ADM).

	Project 1	*Project 2*	*Project 3*	*Project 4*	*Project 5*
Project 1	-	1	0	0	0
Project 2	0	-	1	0	1
Project 3	0	0	-	0	0
Project 4	1	0	1	-	0
Project 5	0	0	1	0	-

In relation to the second phase, which refers to the use of the over-classification or over-classification ratio, the ELECTRE method, through the use of the indices already established above, seeks to obtain a division of the set of alternatives A, since it is a finite set in addition to containing all the alternatives considered, in two subsets N and $A \setminus N$, such that:

- $\forall$ b $\in A$, exist, such that aSb.
- $\forall$ a,b $\in N$, $a\bar{S}b$ and $b\bar{S}a$.
- $N \cap \bar{N}$ is the empty set.
- $N \cup \bar{N}$=A.

3.4.9 ELECTRE graph

If a graphical representation of the relation S is shown, the set N constitutes the kernel of the resulting graph. If the graph has no circuits, the kernel exists and has only a single element. In most cases, the number of kernel alternatives can be reduced, resting the values of $c*$ (decreasing from one), and of $d*$ (increasing from zero).

In the ELECTRE graph, the dominance relationships are represented for each criterion. Each node represents one of the non-dominated alternatives or choices. From node **a** to vertex **b**, a loop is drawn if and only if the corresponding element of the aggregate dominance matrix (ADM) is 1.

For example, if there is a loop oriented from p_1 to p_2, then, it means that no matter what, the criterion p_1 dominates p_2, that is, the alternative p_1 is superior to the alternative p_2. In Figure 3.2 we present the ELECTRE graph for the example provided.

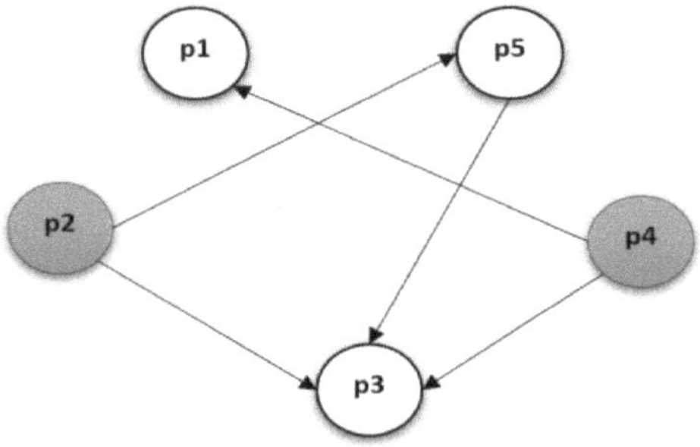

Figure 3.2: ELECTRE graph result of the example.

To detail the Kernel or core behavior of the solution a little more, the total number of times each alternative dominates could be determined and arranged from highest to lowest in such a way that they can be hierarchical, remaining as follows:

3.5 Application Examples

In this section, we will provide real case examples, adapted from consultancy and academic experience that are related to balancing and selecting projects taking into account financial and non-financial criteria. These case studies aim to motivate the reader to use or adapt ELECTRE methods in different contexts.

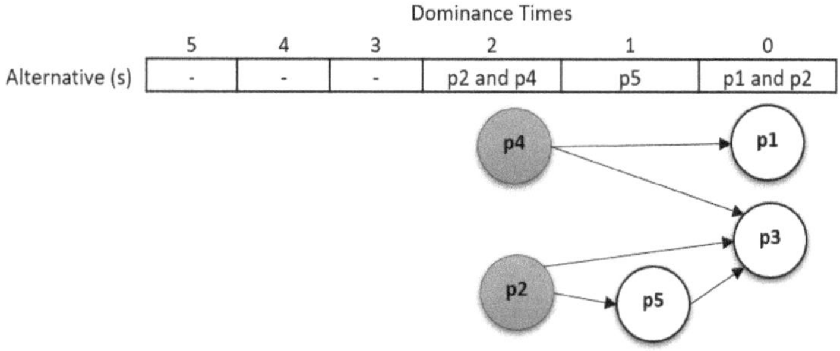

Figure 3.3: ELECTRE graph result reordered by dominance hierarchies.

Case No. 1. Acquisition of specialized software and training package in a financial audit organization.

A financial audit organization with national coverage must acquire specialized software and develop a training program for the members of its work team, in order to increase its portfolio of services and adapt to a new technology that has proven to improve information analysis processes.

The planning team of the organization has identified that there are 4 computer software providers in the market that can meet the expectations of the organization and have undertaken the task of carrying out the respective technical and market studies regarding the following variables (criteria).

Table 3.12: Explanation of the criteria—Case No. 1.

C	Criterion (Decision Variable)	Observations
1	Acquisition cost of acquired software (in Millions)	Licensing value per 5 years
2	Training value per person (in Millions)	Training hour value per person
3	Hours of Training required	Hours of training required
4	Licensing number (access equipment)	Number of computers with software installed
5	Tool usability	Usability (qualitative variable given by the technical study) for the simplicity and ease of use of the application.

To establish a relative importance weight for the identified criteria, the Criteria Comparison Matrix was developed under the AHP based on a work panel made up of the organization's partners and three lead auditors. Results are shown in Table 3.14.

Developing AHP, we were able to identify the following weights of relative importance for the criteria identified, given by Table 3.15.

Table 3.13: Decision matrix of case No. 1.

	C1	*C2*	*C3*	*C4*	*C5*
p1: Supplier A	75	600	55	10	4.5
p2: Supplier B	62	550	40	9	4.0
p3: Supplier C	90	650	25	12	5.0
p4: Supplier D	52	180	45	10	4.0

Table 3.14: Criteria comparison matrix.

	C1	*C2*	*C3*	*C4*	*C5*
C1.Cost	-	4	2	1	2
C2.Training value	1/4	-	2	1/2	2
C3.Training hours	1/2	1/2	-	1/5	2
C4.Number of software licenses	1	2	5	-	5
C5.Usability	1/2	1/2	1/2	1/5	-

Table 3.15: Explanation of the criteria application Case No. 1.

Criterion	W_j
C1.Cost	35.09%
C2.Training Value	8.77%
C3.Training hours	14.04%
C4.Number of Software licenses	35.09%
C5.Usability	7.02%

Solution with ELECTRE I

The ADM for Case No. 1 is given by Table 3.16. With the ADM we can construct the ELECTRE graph showing the dominance over alternatives. The graph solution for Case No. 1 is presented in Figure 3.4, and Figure 3.5 shows the dominance times of each alternative and reorders their hierarchy.

The financial audit organization has one supplier option that dominates the others (p_4 corresponding to Company D). A favorable decision would be made for the organization in the face of a multi-criteria evaluation.

Table 3.16: Aggregate dominance matrix of Case No. 1.

	p1:Supplier A	*p2:Supplier B*	*p3:Supplier C*	*p4:Supplier D*
p1:Supplier A	-	0	1	0
p2:Supplier B	0	-	1	0
p3:Supplier C	0	0	-	0
p4:Supplier D	1	1	1	-

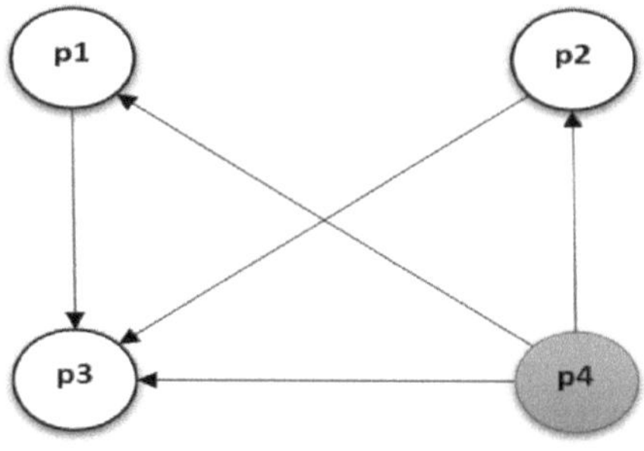

Figure 3.4: Graph solution ELECTRE I Case No. 1.

Dominance Times

	4	3	2	1	0
Alternative (s)	-	p4	-	p1 and p2	p3

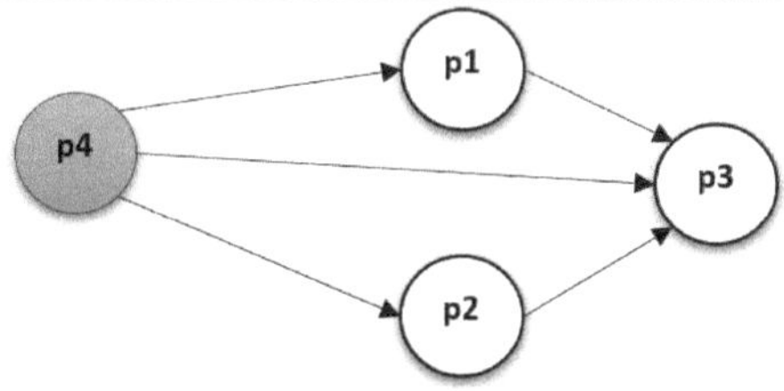

Figure 3.5: ELECTRE graph result reordered by dominance hierarchies of Case No. 1.

Case No. 2. Latin American stock portfolio based on CSR[1] and net income indicators.

The stocks of 50 companies based in Latin America (Colombia, Chile, and Peru,) which had issuance of ADRs[2] on the NYSE stock exchange were analyzed, of which the information report regarding CSR indicators was evaluated as well as the net profit was integrated as a financial criterion.

The selected sample for this case study consists on the 9 companies that presented efficient financial performance and had fully reported the information around their CSR indicators[3].

The Environmental Disclosure Score (EDS), Governance Disclosure Score (GDS), and Social Disclosure Score (SDS) indicators are scores whose maximum value is 100 and would resettle absolute compliance with the Environmental, Governance, and Social parameters based on the Global Reporting Initiative (GRI).

[1]Corporate social responsibility.

[2]American Depositary Receipt (ADR), is a certificate through which investors in the United States of America can trade in shares of listed companies which as companies were incorporated in other countries and not in the United States, being foreign shares with respect to the USA Stock Exchange.

[3]This information is from the year 2017.

Table 3.17: Explanation of the alternatives Case No. 2.

ID	Country	Company name
a:1 CIB	Colombia	Bancolombia
a:2 EC	Colombia	Ecopetrol
a:3 AVAL	Colombia	Grupo Aval
a:4 SCCO	Perú	Southern Copper Corporation
a:5 BVN	Perú	Compañía de Minas Buenaventura S.A
a:6 CCU	Chile	Compañía Cervera unidades S.A
a:7 BSAC	Chile	Banco Santander Chile
a:8 AKOA	Chile	Embotelladora Andina S.A
a:9 EOCCY	Chile	Empresa Nacional de Electricidad S.A

Table 3.18: Explanation of the criteria Case No. 2.

C	Criterion
C1:EDS	Environmental Disclosure Score
C2:GDS	Governance Disclosure Score
C3:SDS	Social Disclosure Score
C4:Net-Inco	Net Income (in Millions of USD)

Table 3.19: Decision matrix of Case No. 2.

	C1:EDS(Max)	*C2:GDS(Max)*	*C3:SDS(Max)*	*C4:Net-Inco(Max)*
W_j	25%	25%	25%	25%
a:1 CIB	46.4286	53.5714	43.3333	$703.14
a:2 EC	63.6364	62.5	54.6875	$1,780.161
a:3 AVAL	2.6786	37.5	18.3333	$583.44
a:4 SCCO	19.3798	55.3571	38.5965	$1,543.00
a:5 BVN	8.5271	26.7857	28.0702	$60.82
a:6 CCU	52.7132	41.0714	47.3684	$211.10
a:7 BSAC	32.1429	33.9286	50.0000	$919.90
a:8 AKOA	42.6357	48.2143	42.1053	$191.93
a:9 EOCCY	32.5581	33.9286	49.1228	$681.59

Solution with ELECTRE I

The ADM of case study No. 2 is presented in Table 3.20. Resulting ELECTRE graphs are shown in Figure 3.6 and Figure 3.7. The latter presents the hierarchical dominance of alternatives.

As can be seen in the results, the kernel of the solution is **a2:EC**, which is the alternative that dominates all of the other alternatives, so it would be the first option among the sample of the nine selected stocks, it would be followed by the alternative **a6:CCU** and **a1:CIB**. On the other hand, the most dominated alternatives are **a5:BVN** and **a3:AVAL** and in the case of alternatives **a7:BSAC** and

Table 3.20: Aggregate dominance matrix of Case No. 2.

	a1	*a2*	*a3*	*a4*	*a5*	*a6*	*a7*	*a8*	*a9*
a:1 CIB	-	0	1	1	1	0	1	1	1
a:2 EC	1	-	1	1	1	1	1	1	1
a:3 AVAL	0	0	-	0	0	0	0	0	0
a:4 SCCO	0	0	1	-	1	0	0	0	0
a:5 BVN	0	0	1	0	-	0	0	0	0
a:6 CCU	1	0	1	1	1	-	1	1	1
a:7 BSAC	0	0	1	0	1	0	-	0	1
a:8 AKOA	0	0	1	1	1	0	1	-	1
a:9 EOCC	0	0	1	0	1	0	1	0	-

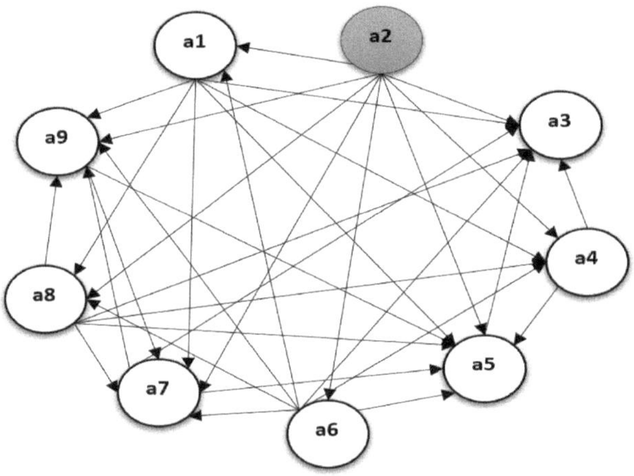

Figure 3.6: Graph solution ELECTRE I Case No. 2.

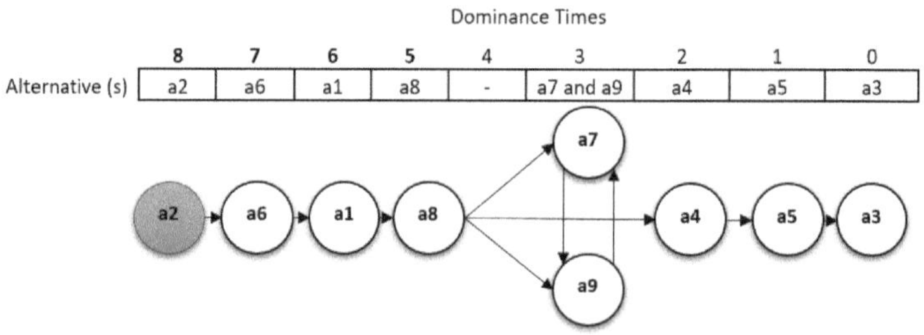

Figure 3.7: ELECTRE graph result reordered by dominance hierarchies of Case No. 2.

a8:AKOA there is a double dominance, which implies an indifference between these two alternatives.

3.6 Conclusions

In this chapter, we show that the application of the ELECTRE I method is a plausible option for solving cases with multiple discrete criteria and alternatives. This method can help a decision-making group focus on the most balanced solution within a group of alternatives, especially those decisions with a high number of alternatives and more than three evaluation criteria. But the main benefit is the combination of quantitative and qualitative criteria from different areas of knowledge such as economic, environmental, technical and human criteria, among others.

In the two cases of applications presented in this chapter, the most dominant options within a group of alternatives were verified. This suggests that these dominant alternatives are the ones to be selected within the possible choices. It is possible that decision-making using ELECTRE I does not present a dominant solution, that is, the calculation did not yield a specific Kernel, this situation indicates that to find the most dominant alternative or alternatives it is necessary to append more criteria with their values in the decision matrix.

References

[1] D.A.G. Chavira, J.C.L. Lopez, J.J.S. Noriega, O.A. Valenzuela and P.A.A. Carrillo. A credit ranking model for a parafinancial company based on the electre-III method and a multiobjective evolutionary algorithm. *Applied Soft Computing*, 60: 190–201, 2017.

[2] H. Costa, L. Nepomuceno and V. Pereira. J-electre-v2.0-user guide. Mar 2019.

[3] M. Doumpos, Y. Marinakis, M. Marinaki and C. Zopounidis. An evolutionary approach to construction of outranking models for multicriteria classification: The case of the electre tri method. *European Journal of Operational Research*, 199(2): 496–505, 2009.

[4] H. Durucasu, A. Aytekin, B. Saraç and E. Orakçi. Current application fields of electre and promethee: A literature review. *Alphanumeric Journal*, 5(2): 229–270, 2017.

[5] J. Figueira, V. Mousseau and B. Roy. *Electre Methods*, pp. 133–153. Springer New York, New York, NY, 2005.

[6] J.R. Figueira, S. Greco, B. Roy and R. Słowiński. Electre methods: Main features and recent developments. In *Handbook of Multicriteria Analysis*, pp. 51–89. Springer, 2010.

[7] J.R. Figueira, V. Mousseau and B. Roy. Electre methods. In *Multiple Criteria Decision Analysis*, pp. 155–185. Springer, 2016.

[8] K. Govindan and M.B. Jepsen. Electre: A comprehensive literature review on methodologies and applications. *European Journal of Operational Research*, 250(1): 1–29, 2016.

[9] A. Guitouni and J.-M. Martel. Tentative guidelines to help choosing an appropriate mcda method. *European Journal of Operational Research*, 109(2): 501–521, 1998.

[10] S.A.S.A. Mary and G. Suganya. Multi-criteria decision making using electre. *Circuits and Systems*, 7(6): 1008–1020, 2016.

[11] G.A. Mendoza and H. Martins. Multi-criteria decision analysis in natural resource management: A critical review of methods and new modelling paradigms. *Forest Ecology and Management*, 230(1-3): 1–22, 2006.

[12] D.L. Olson, A.I. Mechitov and H. Moshkovich. Comparison of mcda paradigms. In *Advances in Decision Analysis*, pp. 105–119. Springer, 1999.

[13] M. Prombo. Software. Package 'OutrankingTools' in R. Software, Dec, 2014.

[14] B. Roy, P. Bertier. La méthode ELECTRE. Note de travail 142. *SEMA-METRA, Metra-International*, 1971.

[15] B. Roy. Classement et choix en pésence de points de vue multiples. *Revue Française D'informatique et de Recherche Opérationnelle*, 2(8): 57–75, 1968.

[16] B. Roy. A multicriteria analysis for trichotomic segmentation problems. *Multiple criteria analysis: Operational methods*, pp. 245–257, 1981.

[17] B. Roy. Paradigms and challenges. In *Multiple Criteria Decision Analysis: State of the Art Surveys*, pp. 3–24. Springer, 2005.

[18] B. Roy. *Multicriteria Methodology for Decision Aiding*, volume 12. Springer Science & Business Media, 2013.

[19] B. Roy et al. Electre III: Un algorithme de classements fondé sur une représentation floue des préférences en présence de critéres multiples. 1978.

[20] B. Roy and J-Chr Hugonnard. Ranking of suburban line extension projects on the paris metro system by a multicriteria method. *Transportation Research Part A: General*, 16(4): 301–312, 1982.

[21] T.L. Saaty. Group decision making and the ahp. In *The Analytic Hierarchy Process*, pp. 59–67. Springer, 1989.

[22] J. Watróbski, J. Jankowski, P. Ziemba, A. Karczmarczyk and M. Zioło. Generalised framework for multi-criteria method selection. *Omega*, 86: 107–124, 2019.

[23] A. Yanie, A. Hasibuan, I. Ishak, M. Marsono, S. Lubis, N. Nurmalini, M. Mesran, S.D. Nasution, R. Rahim, H. Nurdiyanto et al. Web based application for decision support system with electre method. In *Journal of Physics: Conference Series*, volume 1028, p. 012054. IOP Publishing, 2018.

II

OPERATION RESEARCH IN DATA MINING AND CLUSTERING

Chapter 4

Semidefinite Optimization Models in Multivariate Statistics: A Survey

*Eloisa Macedo,** *Tatiana Tchemisova* and *Adelaide Freitas*

4.1 Introduction

Increasing advances in technology have allowed to generate and collect large amounts of data and store large datasets from which it is desirable to extract meaningful information that can be further used to boost knowledge. Popular statistical techniques such as Principal Component Analysis (PCA) and Clustering have been widely used to reduce the dimension of the attribute space and to reveal some patterns hidden among the objects. Clustering and PCA are powerful techniques in Statistics, Data Mining, and Machine Learning with a wide range of applications to data analysis, such as social and psychological fields, computational biology, pattern recognition, and image processing [1, 24].

Dimensionality reduction focuses on representing p-dimensional data in a q-dimensional space with $q < p$. The choice $q = 2$ is often suggested for easier data visualization. For this purpose, PCA finds p linear combinations of p original attributes, called principal components, such that the first q components explain the maximum variability of the p-dimensional data. The orthogonal projection of the data onto a lower-dimensional space provides the best description

Centre for Mechanical Technology and Automation, Dept. Mechanical Engineering, University of Aveiro, Portugal and
Center for Research & Development in Mathematics and Applications, Dept. Mathematics, University of Aveiro, Portugal.
* Corresponding author: macedo@ua.pt

of the original data in the sense that this space captures the highest proportion of total variance [21]. To obtain the coefficients (called unit-loadings) of the principal components which afford the directions of maximum variability of the data, PCA methodology can be performed either by eigenvalue decomposition of the data covariance (correlation, resp.) matrix or through the singular value decomposition of the original (standardized, resp.) data matrix. The number of the returned principal components equals the number of the original attributes. Since the unit-loadings are, in general, nonzero, the interpretation of the principal components can be considered difficult and, consequently, present a shortcoming of the PCA methodology [18, 21].

In an attempt to produce more interpretable principal components, various approaches based on disjoint or sparse PCA-based methodologies, either obtaining principal components in which each attribute contributes to a single principal component (yielding sparse and disjoint components) or turning some loadings to zero (yielding sparse, not necessarily disjoint components), have been proposed in the literature along with associated optimization algorithms. Furthermore, some approaches perform simultaneously clustering of objects (e.g., [1, 14, 23, 38, 44, 51]).

Clustering is an unsupervised statistical technique that aims to group data into a finite number of subsets (clusters). Usually, clustering is referred to objects. Hence, the objects within the same cluster are more similar to one another, than the objects belonging to different clusters. In (partitional) clustering, each cluster can be identified by its centroid, which is a non-observable object representing the mean of all the objects assigned to the cluster [24]. Clustering of attributes (variables), where two attributes belonging to the same cluster are supposed to be a more correlated or associated attribute pair, have been considered in the literature (e.g., [15, 52]). A fundamental key aspect in modeling such problems is to define the (dis)similarity measure.

Among different clustering methods, the most common measures are those that are based on distance functions. In particular, for clustering of objects, the squared Euclidean distance is widely used as a dissimilarity measure. Thus, in such clustering problems, the goal is to minimize the sum of the squared distances between each object and the corresponding clusters' centroid. This problem is often referred to as the minimum sum-of-squares clustering (MSSC) problem and various solution algorithms have been developed to solve it. One of the most popular among them is the k-means algorithm [24].

It turns out that both the PCA and Clustering problems can be modeled in the form of semidefinite problems. Semidefinite programming (SDP) is an important area of convex conic optimization. An SDP problem seeks to minimize (or maximize) a linear objective function subject to constraints in the form of linear matrix inequalities [58]. Being a natural generalization of linear programming (LP), SDP is an active area of research with many applications to combinatorial, robust, and structural optimization, as well as approximation theory, control systems, and data analysis. Semidefinite models arise in mechanical and electrical engineering, computational biology, quantum chemistry, atomic physics, circuit design, sensor network location, and signal processing, among others [3]. Many convex optimization problems can be reformulated as that of SDP and that SDP relaxations of nonconvex optimization problems provide, in general, good approximation bounds on the optimal value.

By reviewing the literature dedicated to the application of semidefinite models to statistical techniques, one can conclude that SDP has great potential for the effective solution of problems related to both PCA and clustering methods. Nevertheless, it seems that the use of SDP models in these two areas is not so common, which may be due to the unfamiliarity of the data analysts community and statisticians with such developments. Thus, the main objectives of this survey are twofold: (1) to make a brief introduction to SDP and the most common solution algorithms, and (2) to review relevant research that focuses on SDP application to statistical techniques, in particular, related to PCA and clustering.

4.2 Semidefinite Programming (SDP)

In SDP, an objective function is minimized under the condition that some matrix-valued function is positive semidefinite. When the objective function is convex and the matrix-valued function is an affine combination of some symmetric matrices, a convex SDP problem is obtained. There are many applications of SDP models to combinatorial optimization, control systems, approximation theory, etc. (see [57, 58] and the references therein).

Linear SDP can be considered as a generalization of LP, where the vector of variables is replaced by a matrix and the constraints have the form of Linear Matrix Inequalities (LMIs). Although there are different similarities between the LP and the SDP, the latter is more general and combines some other problems, such as quadratically constrained problems, second-order cone programming, among others [3]. SDP also presents differences to LP in what concerns duality.

While the LP had a rapid growth between the 1950s and 1960s, mainly due to the emergence of new modeling challenges related to planning and logistics (as a result of the Second World War, the advent of computers, and progress in the development of computational methods), as well as the practical effectiveness of the Simplex Method, SDP has been slower to attract attention, and only around the 90's SDP models became particularly attractive due to the development of effective methods allowing to solve them in polynomial time [3, 50, 58]. Such methods include Interior Point Methods (IPMs) [50], which perform very well in practice, at least for medium-scale problems [3].

4.2.1 SDP problems: basic definitions and duality results

In this chapter, for given integers s, k and t, let $\mathbb{R}^s$ denote the space of s-dimensional real vertices, $\mathbb{R}^{k\times t}$ denote the set of all $(k\times t)$-dimensional matrices, $\mathscr{S}(k)$ stay for the space of symmetric $(k\times k)$-dimensional matrices, and $\mathscr{P}(k)$ be the cone of positive semidefinite symmetric matrices.

Given $\mathbf{B} \in \mathscr{S}(k)$, write $\mathbf{B} \succ 0$ $\mathbf{B}(\succeq 0)$ if $\mathbf{B}$ is positive definite (positive semidefinite), and $\mathbf{B} \prec 0$ $\mathbf{B}(\preceq 0)$ if $\mathbf{B}$ is negative definite (negative semidefinite). Space $\mathscr{S}(k)$ is considered here as a vector space with the trace inner product (denoted by $\operatorname{tr}(\mathbf{AB})$ for given matrices $\mathbf{A}$ and $\mathbf{B}$).

A linear SDP problem is an optimization problem that can be formulated as

$$\begin{aligned} \min_{x\in\mathbb{R}^s} \ & c^\top x \\ \text{s.t. } & \mathscr{A}(x) \preceq 0, \end{aligned} \tag{4.1}$$

where $x \in \mathbb{R}^s$ is a vector variable, $c \in \mathbb{R}^s$ and $\mathscr{A}(x)$ is a matrix-valued function defined as $\mathscr{A}(x) := \sum_{i=1}^{s} \mathbf{A}_i x_i + \mathbf{A}_0$, with $\mathbf{A}_i \in \mathscr{S}(k)$, $i = 0, 1, ..., s$.

It can be assumed, without loss of generality, that the matrices $\mathbf{A}_i$, $i = 1, ..., s$, are linearly independent. If the matrices involved in the SDP problem 4.1 are diagonal, then, this problem reduces to an LP problem.

The Lagrange function corresponding to the SDP problem 4.1 is

$$\mathscr{L}(x,\mathbf{Z}) = c^\top x + \operatorname{tr}(\mathbf{Z}\mathscr{A}(x)) = c^\top x + \operatorname{tr}\left(\mathbf{Z}(\mathbf{A}_0 + \sum_{i=1}^{s} \mathbf{A}_i x_i)\right), \tag{4.2}$$

and the respective dual problem is

$$\begin{aligned} \max_{Z\in\mathscr{S}(k)} \ & \operatorname{tr}(\mathbf{A}_0\mathbf{Z}) \\ \text{s.t. } & -\operatorname{tr}(\mathbf{A}_i\mathbf{Z}) = c_i \ \forall i = 1, \ldots, s, \\ & \mathbf{Z} \succeq 0, \end{aligned} \tag{4.3}$$

where $\mathbf{Z} \in \mathscr{P}(k)$ is the dual matrix variable.

It is well known that the SDP duality results are weaker than that of LP, mainly because the cone of semidefinite matrices is not polyhedral [3].

Denote by $\mathscr{X} := \{x \in \mathbb{R}^s : \mathscr{A}(x) \preceq 0\}$ the feasible set of the SDP problem 4.1, and by $\mathscr{Z} := \{\mathbf{Z} \in \mathscr{P}(k) : \operatorname{tr}(\mathbf{A}_i\mathbf{Z}) = -c_i,\ i = 1, \ldots, s\}$ the feasible set of the corresponding dual problem 4.3. The following property holds in SDP, including lower and upper bounds on the values of the primal and dual SDP problems, respectively.

Theorem 4.1

[Weak Duality Property] Given a pair of feasible solutions $x \in \mathscr{X}$ and $\mathbf{Z} \in \mathscr{Z}$ of the primal and dual SDP problems defined 4.1 and 4.3, respectively, the following inequality holds:

$$c^\top x \geq \mathrm{tr}(\mathbf{A}_0\mathbf{Z}).$$

If for some $x^0 \in \mathscr{X}$ and $\mathbf{Z}^0 \in \mathscr{Z}$, the equality $c^\top x^0 = \mathbf{A}_0\mathbf{Z}^0$ takes a place, then evidently, x^0 is an optimal solution of problem 4.1 and $\mathbf{Z}^0$ is an optimal solution of problem 4.3. Generally, the difference between the optimal values of primal and dual SDP problems is not guaranteed to be zero as it holds in LP. Strong duality, i.e., the property that the optimal values of the problems 4.1 and 4.3 are attained and equal, is ensured only when certain conditions for one of these problems are met [3]. One such condition is the *strict feasibility* (the Slater condition).

Definition 4.1 The constraints of the primal SDP problem 4.1 satisfy the Slater condition if the interior of its feasible set $\mathscr{X}$ is nonempty, i.e.,

$$\exists\, \bar{x} \in \mathbb{R}^s : \mathscr{A}(\bar{x}) \prec 0, \tag{4.4}$$

and the constraints of the dual problem 4.3 satisfy the Slater condition if there exists $\bar{\mathbf{Z}} \succ 0$ such that

$$-\mathrm{tr}\left(\mathbf{A}_i\bar{\mathbf{Z}}\right) = c_i \ \forall i = 1, \ldots, s.$$

SDP problems that satisfy the Slater condition are also called *regular*. Regularity is an important property of optimization problems that is not always verified.

Denote by $v_p \geq -\infty$ the optimal value of the objective function of the primal problem 4.1 and let $v_d \leq +\infty$ stay for the optimal value of the corresponding dual problem 4.3. From Theorem 4.1 it follows that $v_p - v_d \geq 0$.

The following theorem and corollary state that for the regular SDP problems the *duality gap* (the difference between the primal and dual optimal values) vanishes.

Theorem 4.2

[Strong Duality Property] Assume that the optimal value, v_p, of the objective function of the primal problem 4.1 is finite. Under the Slater condition 4.4, the optimal value, v_d, of the corresponding dual problem 4.3 is attained and the duality gap vanishes: $v_p = v_d$.

Corollary 4.1

1. *If the dual SDP problem 4.3 satisfies the Slater condition with* v_d *finite, then,* $v_p = v_d$ *and this value is attained for the primal problem 4.1.*
2. *If both primal and dual SDP problems satisfy the Slater condition, then,* $v_p = v_d$ *and this value is attained for both problems.*

4.2.2 SDP solvers

Various numerical methods have been developed for solving SDP problems. Among them, the (primal-dual) IPMs are considered to be the most efficient [3]. These methods are based on the assumption that the Slater condition is satisfied, and in the attempt to solve both primal and dual SDP problems, to minimize the duality gap. The basic idea of an IPM is to replace the solution of problem 4.1 with a sequence of approximated solutions of some (perturbed) auxiliary problem to avoid traversing the cone during the optimization process.

Among general-purpose publicly available SDP solvers that implement IPMs running in finite precision, the most popular are CSDP, SDPA, DSDP, SDPT3, and SeDuMi (for the references see [3]), many of them being available in the R programming language or MATLAB®. Most of these solvers find a solution to an SDP problem to arbitrary accuracy in a polynomial time, but there is a backbone related to dimensionality since such solvers may run out of memory while forming and solving large linear systems. Thus, the development of specially built methods for handling large-scale SDP problems, either based on augmented Lagrangian alternating direction method or on low-rank solutions, has become an active field of research (for details, see the recent survey [34]). A first-order algorithm suitable to solve many optimization problems, including SDP, was proposed in [40] and its implementation in the Splitting Conic Solver showed to be quite efficient in handling large-scale optimization problems.

A particular issue in SDP numerical solution is related to the basic assumption of the problem's regularity since the strong duality condition is only ensured if the SDP problem is regular. There exist many SDP problems arising from different applications that fail regularity and researchers have been addressing this in their studies (e.g., [9, 33, 41, 53, 54]). Popular solvers may run into numerical difficulties in case of nonregularity, possibly returning wrong or infeasible solutions.

In [30], a new approach to verifying the problem's regularity before proceeding with its solution is proposed. The approach is based on a specific concept of immobile indices of constraints. Based on this concept, an algorithm named DIIS (*Determination of the Immobile Index Subspace*) was developed in [25] for finding a basis of the immobile index subspace. It was shown that the dimension of the immobile index subspace is an important characteristic of the feasible set of an SDP problem and the problem is regular in terms of the Slater condition if and only if the subspace of immobile indices is null. In [30], it was presented a numerical procedure that permits to test regularity of a given SDP problem in a finite number of steps. This procedure can be used as a tool to guarantee trustworthiness in the results obtained by any SDP solver.

In [9], a preprocessing procedure for regularization of abstract convex and conic problems for which any regularity conditions fail, was suggested. This procedure is based on the concept of the minimal face of the problem's constraint cone. If a given problem fails to satisfy the regularity in the form of the Slater condition, the preprocessing procedure transforms this problem into an equivalent form where the constraints cone is substituted by its minimal face. Being applied to SDP, the regularization procedure is based on the Borwein-Wolkowicz *Facial Reduction Algorithm* (FRA) [7]. Although the FRA algorithm permits regularizing a given SDP problem into an equivalent regular one in one step, this algorithm is not explicitly described.

Only around 2014 some algorithmic implementations of facial reduction procedures became available. More recently, an exact algorithm based on symbolic homotopy for solving SDP problems without assumptions on the feasible set was proposed in [19]; however, the algorithm is only suitable for small-scale SDP problems. These issues have motivated the development of libraries of SDP instances of particular structures that are considered essential for developing and testing new stopping criteria for SDP methods, and conceiving more efficient solvers (e.g., [33, 54]).

4.3 SDP Application to Multivariate Statistical Techniques

This section summarizes relevant articles on models and methods applied to statistical methods based on SDP. The SCOPUS database was selected for search and collecting the presented references under the two main research streams of multivariate statistical techniques, mainly PCA and Clustering and from a practical perspective applied in data analysis. A systematic approach was performed by exploring publications within the period from January 2001 to January 2021 by searching strings in TITLE-ABS-KEY which contain the terms ((semidefinite OR SDP OR SDO) AND (model OR problem) AND (data analysis OR data mining OR statistic)) AND (Principal Component Analysis OR PCA OR clustering OR dimensionality reduction OR high-dimension). This search resulted in a list of more than 90 documents that were narrowed by considering only relevant documents written in English and addressing the explicit use of SDP formulation or relaxation on data analysis problems. It is worth mentioning that besides Mathematics and Computer Science, it was found a variety of SDP application in areas such as engineering, neuroscience, energy, chemistry and economics, econometrics, and finance.

These publications and other equally relevant e-prints available in the open-access archive were further studied, resulting in the list of selected papers presented here. The following will be a discussion of the selected relevant references.

4.3.1 Principal component analysis

In defining new variables, i.e., principal components that are linear functions of the correlated original variables in a given dataset, the PCA technique allows the reduction of dimensionality while preserving as much variability, i.e., statistical information, as possible. The establishment of these components is obtained by solving an eigenvectors/eigenvalues problem of the data covariance matrix [22]. More concretely, given an $(m \times n)$-dimensional centered data matrix $\mathbf{D}$, the coordinates of the q-th principal component $v^q \in \mathbb{R}^n$, i.e., unit-loadings, are found by solving the following problem [21]:

$$\begin{aligned} \max_{v^q \in \mathbb{R}^n} \quad & (v^q)^\top \Sigma v^q \\ \text{s.t.} \quad & (v^q)^\top v^q = 1, \\ & (v^j)^\top v^q = 0, \ j = 1, 2, \cdots, q-1, \end{aligned} \tag{4.5}$$

where the matrix $\Sigma \in \mathbb{R}^{n \times n}$ is proportional to the matrix $\mathbf{D}^\top \mathbf{D}$.

Hence, the first q principal component loadings also render an optimal solution to the following problem [45]:

$$\begin{aligned} \max_{\mathbf{V} \in \mathbb{R}^{n \times q}} \quad & \operatorname{tr}\left(\mathbf{V}^\top \Sigma \mathbf{V}\right) \\ \text{s.t.} \quad & \mathbf{V}^\top \mathbf{V} = \mathbf{I}_q, \end{aligned}$$

where Σ is assumed to be a covariance matrix, without loss of generality, $\mathbf{V} := [v^1 \cdots v^q]$ represents the $(n \times q)$-dimensional matrix composed by the units-loadings of the first q principal components, and $\mathbf{I}_q$ is the $(q \times q)$-dimensional identity matrix.

Furthermore, in the PCA problem's formulation 4.5, a sparsity criterion based on the cardinality (number of non-zero elements or 0-norm) of the $(n \times 1)$ vector v^q can be inserted leading to a so-called *sparse PCA problem*. Denoting by x the solution-vector v^1 related to the first sparse principal component, the sparse PCA problem can be generally formulated as

$$\begin{aligned} \max_{x \in \mathbb{R}^n} \quad & x^\top \Sigma x \\ \text{s.t.} \quad & ||x||_2 = 1, \\ & ||x||_0 \leq \ell, \end{aligned} \tag{4.6}$$

where $\ell \leq n$ is an integer parameter controlling the sparsity of the solution [10]. Due to the sparsity constraint, problem 4.6 is well-known to be *NP-hard*.

In [11], the sparse PCA problem 4.6 was rewritten in terms of a rank-one symmetric $(n \times n)$-dimensional matrix $\mathbf{X} = [x_{ij}]$ which can be written in the form $\mathbf{X} = xx^\top$. The authors of [11] have proposed an SDP relaxation for this problem, in which an optimal solution will be the upper bound on the optimal value of the sparse PCA problem.

The ultimate SDP relaxation is given by

$$\begin{aligned} \max_{\mathbf{X} \in \mathscr{S}(n)} \quad & \operatorname{tr}(\Sigma \mathbf{X}) \\ \text{s.t.} \quad & \operatorname{tr}(\mathbf{X}) = 1, \\ & \mathbf{1}^\top |\mathbf{X}| \mathbf{1} \leq \ell, \\ & \mathbf{X} \succeq 0, \end{aligned} \tag{4.7}$$

where $|\mathbf{X}|$ corresponds to the matrix whose elements are the absolute values of the elements of $\mathbf{X}$ and $\mathbf{1}$ represents the n-vector of ones.

An extension to nonsquare matrices was also suggested in [11]. In particular, the authors showed that the proposed SDP relaxation could be a robust formulation of the maximum eigenvalue problem. For small-scale problems, these formulations can be efficiently solved using IPMs, but for large-scale instances, it was proposed to explore the problem's structure and a first-order algorithm for solving the semidefinite formulation was developed. A specific application to the sparse PCA was described, and several numerical experiments were presented comparing the proposed approach with other modified PCA techniques such as PCA with simple thresholding and PCA using the LASSO approach (SCoTLASS).

More recently, in [10], an algorithm based on an SDP relaxation of the sparse PCA problem 4.6 with the constraint $||x||_2 = 1$ replaced by $||x||_2 \leq 1$, which does not assume a specific model for the covariance matrix, was suggested. Specifically, the procedure involves a two-step rounding scheme that converts a solution matrix into a sparse vector. The SDP relaxation of the problem 4.6 was then defined as follows:

$$\begin{aligned} \max_{\mathbf{X} \in \mathscr{P}(n)} \quad & \operatorname{tr}(\Sigma \mathbf{X}) \\ \text{s.t.} \quad & \operatorname{tr}(\mathbf{X}) \leq 1, \\ & \sum_{i,j} |x_{ij}| \leq \ell. \end{aligned} \tag{4.8}$$

Notice that the SDP model 4.8 is written in its dual form. It is shown in [10] that a solution to this problem is a positive semidefinite matrix $\mathbf{X}^*$ that is an upper bound of the true optimal solution of the sparse PCA in Equation 4.6. Generating a random Gaussian vector, an unbiased estimator for the trace of $\mathbf{X}^{*\top} \Sigma \mathbf{X}^*$

can be constructed. By using the von Neumann trace inequality and a sparsification procedure that guarantees that the larger entries are kept, while the smaller ones are set equal to zero, it is possible to obtain a sparse vector. To increase the probability of success, multiple Gaussian vectors g should be generated, and only the one maximizing $g^\top \mathbf{X}^{*\top} \Sigma \mathbf{X}^* g$ should be chosen. Experiments with gene expression and text classification datasets demonstrated the efficiency of the proposed algorithm in finding a sparse solution. The performance of the SDP-based algorithm was shown to be comparable to other methods, in particular, for more sparse output vectors, this algorithm performed much better than other methods for sparse PCA (e.g., R-function `spca`), yielding higher levels of accuracy, but compromising running time, though in a low-degree polynomial time.

In [12], a new SDP relaxation to the problem of finding sparse factors explaining a maximum amount of variance providing upper bounds, was formulated and sufficient conditions for the global optimality were derived. The authors also developed a greedy algorithm to solve the sparse PCA problem, however, some limitations in terms of dealing with large instances were reported. Later, using randomization techniques, the authors of [13] presented a study on approximation bounds for an SDP relaxation of the sparse eigenvalue problem. Following the relaxation suggested in [11], new bounds were derived which depend on the optimum value of the SDP relaxation in Equation 4.8. Concretely, it was shown that the higher the optimum value, the better the approximation. The authors of [13] regularized the relaxation and used a conditional gradient algorithm to solve it. Numerical results revealed better performance in solving some problems when compared to other methods.

It is worth mentioning that the interesting developments in [11] received a lot of attention from many scientific studies. For instance, in [2], based on the SDP relaxation of the sparse eigenvector problem from [11], the authors considered a spiked model for the covariance, derived explicit sample size thresholds for recovery of the true sparse vectors, and showed that the SDP relaxation under some conditions has a unique rank-one solution and that the high-dimensional data has greater statistical efficiency. The obtained results (also using information-theoretic methods) revealed the existence of trade-offs between statistical and computational efficiency in high-dimensional eigen-analysis. In [27], it was reported a numerical implementation for the optimization algorithm suggested in [11]. The authors showed that using only a partial eigenvalue decomposition on the current iterate, a sufficiently precise gradient approximation can be produced, improving computational efficiency.

Numerical experiments on gene expression datasets used sparse PCA as a tool to produce clusters of variables and were able to group more relevant genes than other methods. In [36], two approaches for robust PCA based on SDP were studied. The first one is based on a modification of PCA with a function that is less sensitive to outliers called the mean absolute deviation. This yielded a nonconvex problem which was reformulated in the form of an SDP model and easily solved applying a rounding procedure. The second approach is based on a low-leverage decomposition of data, where the data is represented in the form of two matrices after having separated the corrupted observations. The proposed algorithm achieved the direction of maximum deviation and was revealed to be quite efficient in obtaining solutions. A faster block coordinate ascent algorithm that can be used to solve the SDP relaxation from [11] is described in [60]. The procedure starts by reducing the above mentioned SDP relaxation by applying a safe feature elimination method before solving it. Numerical experiments on real-world test collections show that this approach can be applied to very large-scale data sets.

Following the convex relaxation of the sparse PCA problem proposed in [11], Shiqian Ma in [28] developed an alternating direction method based on a variable-splitting technique and an augmented Lagrangian framework. An interesting feature is that it was proposed to solve directly and exactly the primal problem instead of the dual one as in [11]. This change guarantees that the resulting matrix is sparse. Global convergence results were established and numerical experiments showed that the method is quite effective.

The authors of [59] gathered together a survey on various convex relaxations which provided good approximate solutions for sparse PCA, as well as algorithms for solving these relaxations.

In [20], the tensor-based PCA problem was studied. A tensor is a multidimensional generalization of matrices which has many important applications in image processing, statistical learning, or bioinformatics. Tensor-based PCA problems can be modeled as SDP instances by relaxing the rank-one constraint since in general, the tensor can be reduced and embedded into a symmetric matrix. An alternating direction method of multipliers was used to solve the resulting SDP models. In the case when no rank-one solution is obtained for the SDP relaxation, a small perturbation may be applied to the original tensor, or a post-processing procedure may be performed. Numerical results showed that the suggested approach can provide optimal solutions for the original tensor PCA problem quite efficiently.

In [39], the authors proposed a quantitative definition of the accuracy of the dimensionality reduction (note that a reduction is considered to be accurate if the information about a protected class cannot be inferred from the dimensionality-reduced data point) and formulated an SDP model based on mean and covariance constraints for FPCA (Fair PCA). The proposed approach was focused on extracting the d largest eigenvectors from the optimal solution of the SDP relaxation model. Numerical results highlighted its capabilities and a particular application to health data showed that the use of FPCA is also suitable to perform a fair clustering.

It was recently proposed in [5] an exact reformulation of sparse PCA as a convex mixed-integer SDP problem. A cutting-plane method was developed to solve this problem to certifiable optimality, providing small bound gaps at large-scale problems. A randomized rounding procedure was also developed and the proposed relaxation was compared to that formulated in [11]. The application of the proposed approach showed to be quite efficient in providing interpretable principal components for financial and medical datasets. Some extensions were also discussed, in particular related to the possible development of a certifiably optimal algorithm for non-negative sparse PCA.

A novel approach for robust PCA (RPCA) with partially observed data was recently suggested in [29]. It consists of constructing a nonconvex and nonsmooth SDP reformulation of the original NP-hard RPCA model by adding a redundant semidefinite cone constraint and solving small subproblems using a proximal alternating linearization method that was proposed in [6]. Each subproblem results in an exposing vector for a facial reduction technique that is able to reduce the size significantly. The approach takes into account the low-rank and sparse structure and yields highly accurate solutions. The reported numerical experiments indicate great capability to recover large-scale problems' solutions.

4.3.2 Clustering

Clustering is an unsupervised statistical technique that is very useful for data analysis in a wide range of applications from the engineering or medical science to the humanities. It groups objects or variables based on some inherent dissimilarity/similarity measure among them, being the most common dissimilarity measure between objects the Euclidean distance (lower the distance, lower the dissimilarity), and the most common similarity measure between variables is the Pearson correlation (higher the correlation, higher the similarity) [14]. The most popular model for the clustering problem is the minimum sum-of-squares clustering (MSSC) one that can be formulated concerning a clustering of m objects into $p < m$ clusters (without loss of generality), as follows [24, 42]:

$$\begin{array}{ll} \min\limits_{u_{ij}} & \sum\limits_{j=1}^{p}\sum\limits_{i=1}^{m} u_{ij} \left\| d^i - \frac{\sum\limits_{t=1}^{m} u_{tj} d^t}{\sum\limits_{t=1}^{m} u_{tj}} \right\|_2^2 \\ \text{s.t.} & \sum\limits_{j=1}^{p} u_{ij} = 1, \ i = 1,\dots,m, \\ & \sum\limits_{i=1}^{m} u_{ij} \geq 1, \ j = 1,\dots,p, \\ & u_{ij} \in \{0,1\}, \ i = 1,2,\dots,m, \ j = 1,\dots,p, \end{array} \tag{4.9}$$

where $\mathbf{D} = [d_{ij}]$ is the $(m \times n)$-dimensional data matrix considered centered, without loss of generality, $d^i \in \mathbb{R}^n$ represents the i-th row of the matrix $\mathbf{D}$, and $\mathbf{U} = [u_{ij}]$ is a $(m \times p)$-dimensional binary and row stochastic matrix representing the object assignment matrix. The first constraint ensures that each object is assigned to a single cluster and the second that each cluster has at least one object assigned. Any feasible solution of the problem 4.9 is an assignment matrix $\mathbf{U}$.

An interesting application of nonlinear SDP models to clustering problems was proposed in [26] and [42]. These works are mainly focused on considering that the MSSC problem 4.9 is usually formulated as a binary integer programming problem, which can be rewritten as an (0,1)-SDP problem. In [42], it is shown that the objective function in the MSSC problem is equal to the function $\operatorname{tr}\left(\mathbf{D}\mathbf{D}^\top(\mathbf{I}_m - \mathbf{Z})\right)$, where the variable $\mathbf{Z}$ is an $(m \times m)$-dimensional matrix, and the SDP formulation of this problem can be given as

$$\begin{aligned} \max_{\mathbf{Z}} \quad & \operatorname{tr}(\mathbf{D}\mathbf{D}^\top\mathbf{Z}) \\ \text{s.t.} \quad & \mathbf{Z}^\top = \mathbf{Z}, \\ & \mathbf{Z}^2 = \mathbf{Z}, \\ & \operatorname{tr}(\mathbf{Z}) = p, \\ & \mathbf{Z}\mathbf{1} = \mathbf{1}, \\ & z_{ij} \geq 0 \qquad \forall i,j = 1,2,...,m. \end{aligned} \tag{4.10}$$

In [42], the authors presented also an approximation algorithm for solving such models that first, focuses on obtaining linear SDP relaxations and their solutions and then, on a rounding procedure that is based on the application of a heuristic to obtain a feasible solution for the nonlinear model. Numerical experiments highlight that this approach provides very efficient procedures especially in the case of two clusters. In [26], following the nonlinear SDP model from [42], it was proposed to obtain a solution for the clustering problem by considering a low-rank SDP model (that in turn does not represent a convex relaxation) and to solve it using a nonconvex optimization algorithm. The authors showed that the problem 4.10 is equivalent to

$$\begin{aligned} \max_{\mathbf{Z}} \quad & \operatorname{tr}(\mathbf{D}\mathbf{D}^\top\mathbf{Z}) \\ \text{s.t.} \quad & \operatorname{tr}(\mathbf{Z}) = p, \\ & \mathbf{Z}\mathbf{1} = \mathbf{1}, \\ & \operatorname{rank}(\mathbf{Z}) = p, \\ & \mathbf{Z} \succeq 0, \\ & z_{ij} \geq 0 \qquad \forall i,j = 1,2,...,m. \end{aligned} \tag{4.11}$$

This SDP problem is solved exactly, and no rounding procedure is applied to obtain a discrete clustering of the data. The suggested algorithm uses low-rank factorizations to optimize full-rank SDP problems, and all needed computations are done in the low-rank space, which leads to improvements in terms of running time and memory. The application of this approach to many datasets provided good results which, in some cases, outperform other SDP approaches.

Following [42], interesting results were obtained in [4]. The authors proposed another SDP relaxation of the k-means problem 4.9, where the optimal solution can be recovered directly (without any rounding step). The authors referred to this occurrence as to *exact recovery*, *tightness*, or *integrality*. The resulting SDP relaxation of problem 4.9 is complete at a significantly closer range, meaning that it is possible to recover the global k-means optimal solutions if the underlying data can be partitioned in unit-radius clusters whose centers are separated by a significantly small distance. It was proved that the proposed SDP relaxation recovers clusters as the unique optimal solution with high probability.

In [47], an SDP-based model was proposed for ensemble clustering, consisting of aggregating multiple solutions into one that maximizes the agreement in the input ensemble. The proposed formulation is an $0-1$ SDP problem that is further relaxed to a polynomial-time-solvable SDP problem, also constructed using the ideas from [42]. After solving the relaxed SDP problem, a rounding procedure based on a variant of the *winner-takes-all* approach is applied (i.e., the singular value decomposition is performed) to obtain column vectors, that may form a matrix whose largest entry in each row will be rounded to 1 and the remaining entries rounded to 0. Numerical results showed promising results that turn to reveal good ensemble quality, in which the major limitation is the need to solve the SDP relaxation problem.

In graph clustering problems, common in the image analysis where a complete graph with positive (similar) and negative (different) edges is considered, the goal is to construct groups with the maximal number of agreement edges within clusters (i.e., it minimizes the number of disagreements). The number of clusters does not need to be specified. This problem is known to be NP-hard [49]. Various approximations and lower bounds were proposed in the literature for minimizing disagreements and maximizing agreements for general weighted graphs. One of them is due to Swamy [49], who developed an approximation algorithm with a 0.7666 performance ratio based on an SDP relaxation of the problem, followed by a rounding procedure for its optimal solution.

In [17], the authors focused on the stochastic block model for the graph partitioning problem and showed that the SDP relaxations can guarantee exact recovery for the general case of a fixed number of clusters, even in the presence of outliers or without the knowledge of the cluster sizes. The approach is based on a penalized SDP and a sufficient condition for exact recovery with high probability is established.

Considering biclustering, mainly applied in genomic contexts where objects (tissue samples) and features (genes) are simultaneously sought according to their expression levels, Ames [1] presented an approach that yields an SDP relaxation with similar recovery guaranteed for the biclustering problem. There were also established conditions that assure that the clusters corresponding to the optimal solution of the SDP problem can be recovered with high probability. These results were recently strengthened by an application to the densest disjoint clique problem consisting of establishing the smaller clusters which can be recovered in the presence of approximately sparse noise [43]. It was shown that the clusters can be recovered from a solution of an SDP relaxation with a high probability if the input graph is sampled from a distribution of a clusterable graph.

In [48], the authors addressed the problem of noisy similarity information on a clustering problem. Based on the SDP formulation suggested in [1], they proposed a way of exact recovering the clusters and detecting the outliers, in which the difference lies in not considering the prior information about the number of clusters. Conditions for exact recovery of the clusters in the presence of outliers were established, mainly focused on certain requirements of the similarity matrix. Numerical experiments on both balanced and unbalanced clusters were presented supporting the conclusion about the effectiveness of this approach.

Very recently, a new model for variable clustering was proposed in [8]. The authors, motivated by the relaxation presented in [42], developed the PECOK (Penalized convex k-means) procedure that uses SDP on a k-means fashion for variable clustering that showed to be generally applicable and guarantees the exact recovery of minimally separated clusters. The procedure focuses on the primal problem and combines different duality-norm bounds. Interesting results were presented for G-block covariance models constructed for clustering of brain regions. In [46], a clustering problem was handled through a flexible probabilistic model and motivated by the work of [42]. The resulting exact clustering with high probability was solved using an SDP estimator that can be considered a relaxed version of k-means. The authors adapted the PECOK algorithm to the clustering problem of a mixture of sub-Gaussian distributions and showed that the obtained results are optimal or near-optimal in recovering the partition and flexible in terms of the number of unknown groups. Numerical experiments also revealed that the implementation of a bias-removing correction step improves the solution's recovery and is an effective approach for handling high-dimensional data.

The authors of [37], inspired by existing work in convex relaxations for clustering problems, obtained new results in the case of the k-means loss by using the SDP relaxations of the k-means clustering problem. In particular, an SDP relaxation of the sublevel set problem was obtained by introducing an inequality constraint to the formulation from [4]. Sufficient conditions guaranteeing near-optimality and uniqueness (up to small perturbations) for the clustering model were deduced.

In many real-world datasets, clusters can overlap and in [55], it was proposed a NEO-k-Means (Non-Exhaustive, Overlapping k-Means) objective, that can be regarded as a reformulation of the k-means problem 4.9 where clusters need not be disjoint, that is, $\mathbf{U}$ may have multiple ones in a row (in the case when a data point belongs to more than one cluster), and/or have zero-rows (in the case when a data point does not belong to any cluster). Thus, specific constraints were added to the model. To handle the resulting problem, an SDP formulation was proposed, mostly due to the need to provide a good initialization for the NEO-k-Means algorithm developed by the authors. A mixed-integer, rank-constrained SDP problem was developed and further relaxed. For this purpose, it was be defined a (positive semidefinite) co-occurrence matrix from an assignment matrix **U**, in which each entry is non-zero if items co-occur in a cluster. Then, the data matrix written in the form of a kernel matrix of the data points was considered. The rank constraint was also dropped and discrete constraints were relaxed. The authors developed a fast iterative algorithm based on low-rank SDP techniques, in particular, they proposed a solution procedure using an augmented Lagrangian method. Numerical results showed that this approach is effective in finding ground-truth clusterings that have varied overlap and non-exhaustiveness.

4.3.3 Clustering and disjoint principal component analysis

When facing many observations described by many variables, it is useful to consider both the dimensionality reduction in terms of the attribute space and the clustering of objects. There exist approaches that find subsets of variables and objects defining homogeneous blocks by using double clustering techniques. Double (rather than a sequential) clustering is particularly valuable since it allows to reveal eventual interactions or dependencies between specific objects and variables, which in turn may help in data interpretation (e.g., [35]). For instance, in [56], a sparse k-means clustering was proposed for identifying groups of objects applying the k-means method and to find variables with the most influence in this classification.

Using a different approach but with an analogous purpose, in [51] a constrained PCA methodology, called *Clustering and Disjoint Principal Component Analysis* (CDPCA) was proposed to simultaneously identify clusters of objects and find a partitioning of variables such that the between cluster deviance in the reduced space of such partition is maximized. This approach is of particular interest and has attracted more attention (e.g., [16, 32, 38, 44]), because it presents a significant advantage over common PCA: CDPCA provides a model consisting of disjoint components, which may be helpful for interpretation purposes.

A quadratic mixed integer program can be considered for CDPCA as

$$
\begin{array}{lll}
\max\limits_{\mathbf{U},\mathbf{V},\mathbf{A}} & \|\mathbf{U}\bar{\mathbf{D}}\mathbf{A}\|_2^2 & \\
\text{s.t.} & u_{ij} \in \{0,\ 1\}, & i=1,\ldots,m,\ j=1,\ldots,p; \\
& \sum\limits_{j=1}^{p} u_{ij} = 1, & i=1,\ldots,m; \\
& v_{ij} \in \{0,\ 1\}, & i=1,\ldots,n,\ j=1,\ldots,q; \\
& \sum\limits_{j=1}^{q} v_{ij} = 1, & i=1,\ldots,n; \\
& \sum\limits_{i=1}^{n} a_{ij}^2 = 1, & j=1,\ldots,q; \\
& \sum\limits_{i=1}^{n} (a_{ij}a_{ir})^2 = 0, & j=1,\ldots,q-1,\ r=j+1,\ldots,q,
\end{array}
\tag{4.12}
$$

where $\mathbf{U} = [u_{ij}]$ ($\mathbf{V} = [v_{ij}]$) is an $(m \times p)$ ($(n \times q)$)-dimensional binary and row stochastic matrix representing the object (variable) assignment matrix, $\mathbf{A} = [a_{ij}]$ is the $(n \times q)$-dimensional component loading matrix, and $\bar{\mathbf{D}}$ is the $(p \times n)$-dimensional object cluster centroid matrix, whose rows correspond to p clusters presented by their centroids which were obtained from the rows of the $(m \times n)$ data matrix $\mathbf{D}$.

The first two constraints in the problem 4.12 correspond to the assignment of each of m objects into only one of p non-empty clusters and the next two constraints represent the assignment of n attributes into q disjoint components. The remaining constraints are associated with a standard PCA modification in order to determine disjoint components, and thus construct a column-wise orthonormal matrix $\mathbf{A}$ where, for any two different columns, one of the corresponding entries is zero, and all the original attributes are included in the new components. The objective function value given by $||\mathbf{U}\bar{\mathbf{D}}\mathbf{A}||_2^2$ can be computed either by $\mathrm{tr}\left(\mathbf{U}\bar{\mathbf{D}}\mathbf{A}(\mathbf{U}\bar{\mathbf{D}}\mathbf{A})^\top\right)$, corresponding to the between-cluster-deviance, or by $\mathrm{tr}\left((\mathbf{U}\bar{\mathbf{D}}\mathbf{A})^\top \mathbf{U}\bar{\mathbf{D}}\mathbf{A}\right)$, representing the total variance of the (centered) data in the reduced space, where the objects are identified by their centroids.

Since this optimization problem is difficult to solve due to the presence of discrete variables and in general, its solving involves heuristic procedures. In [32], a detailed two-step-based scheme of an alternating least-squares (ALS) algorithm, proposed initially in [51] for estimating the parameters of the CDPCA model, was described. It mainly involves the assignment of objects using k-means and the reduction of the attribute space applying PCA to the resulting centroids. In the simplified version suggested in [32], the matrices $\mathbf{V}$ and $\mathbf{A}$ are sequentially constructed by an iterative procedure applied to their rows and columns, to maximize the between cluster deviance in the reduced space, while in the ALS algorithm the matrices are updated separately.

A new approach to solving the CDPCA problem 4.12 by combining SDP-based clustering models and the CDPCA model to maximize the between-cluster-deviance in the reduced space of the components, was proposed in [31], motivated by the works [32, 42] and [51]. It was called the *Two-Step-SDP* algorithm and the core idea is to consider SDP models of the form 4.10 suggested by [42] for clustering both objects and attributes. More concretely, two clustering problems reformulated as an SDP-based model in the form 4.10 were considered in [31]: a k-means clustering problem for grouping objects and a k-means clustering problem for grouping variables. To solve them, a couple of modifications in the ALS algorithm from [32, 51] was introduced, in particular, an application of the SDP-based approximation algorithm from [32, 42] to construct the initial matrices $\mathbf{U}$ and $\mathbf{V}$ (instead of a random choice), and updating the component loading matrix $\mathbf{A}$ using the current assignment matrix $\mathbf{V}$. Based on the iterative procedure for clustering objects described in [42], both clustering problems (one related to objects and another to variables) are solved by first solving approximately the SDA-based model 4.10 associated with the objects (variables) clustering in the form that uses the computation of the first $p-1$ $(q-1)$ eigenvalues in the decreasing order and obtains an approximate solution of the relaxed problem. Then, the second step involves a rounding procedure that applies PCA methodology to find the component loading matrix, the component score matrix, and the object centroid matrix in the reduced space, where the k-means algorithm is performed in the reduced space of the components.

The modifications proposed in [31] were expected to provide improvements, at least in terms of computational time. A recent empirical comparative study of both the ALS algorithm and the Two-Step-SDP algorithm on the estimation of the parameters of the PCA model was executed in [16]. The authors found that the Two-Step-SDP algorithm provides faster results than the ALS algorithm [32, 51], it outperformed ALS in recovering the true variable partitioning unveiled by the datasets and also the true object clusters for data with more structure complexity, and provided higher variance in the case of datasets with two object clusters.

4.4 Conclusions

The diversity of procedures described and discussed in this survey highlights the potential ability to use SDP models in multivariate statistical methods, in particular in the fields related to PCA, providing a tool to reduce the dataset along the direction of maximum variance with minimum information loss and to apply clustering techniques, which focus on grouping objects/variables based on some dissimilarity/similarity measures. According to the presented discussion, SDP has shown to be an important tool enabling robust estimates in several statistical methodologies, but data analysts seem to be unfamiliar with this, maybe due to the lack of more applied research highlighting all SDP capabilities in revealing hidden meaningful information on data.

Acknowledgements

The work of the first author was supported by UIDB/00481/2020 and UIDP/00481/2020-FCT; CENTRO-01- 0145-FEDER-022083; InFLOWence (POCI-01-0145-FEDER-029679), co-funded by COMPETE2020, Portugal2020 - Operational Program for Competitiveness and Internationalization (POCI), European Union's ERDF (European Regional Development Fund) and FCT; and Interreg Europe Project PriMaaS PGI05830, and the work of the second and third authors was supported by the Center for Research and Development in Mathematics and Applications (CIDMA, University of Aveiro) through the Portuguese Foundation for Science and Technology (FCT - Fundação para a Ciência e a Tecnologia), project UIDB/04106/2020 and UIDP/04106/2020.

References

[1] B.P.W. Ames. Guaranteed clustering and biclustering via semidefinite programming. *Mathematical Programming*, 147(1): 429–465, 2014.

[2] A.A. Amini and M.J. Wainwright. High-dimensional analysis of semidefinite relaxations for sparse principal components. *Annals of Statistics*, 37(5B): 2877–2921, 2009.

[3] M.F. Anjos and J.B. Lasserre. (eds.). *Handbook of Semidefinite, Conic and Polynomial Optimization: Theory, Algorithms, Software and Applications, International Series in Operational Research and Management Science*, volume 166. Springer, 2012.

[4] P. Awasthi, A.S. Bandeira, M. Charikar, R. Krishnaswamy, S. Villar and R. Ward. Relax, no need to round: Integrality of clustering formulations. In *Proceedings of the 2015 Conference on Innovations in Theoretical Computer Science*, pp. 191–200. Association for Computing Machinery, 2015.

[5] D. Bertsimas, R. Cory-Wright and J. Pauphilet. Solving large-scale sparse PCA to certifiable (near) optimality. *Journal of Machine Learning Research*, 1: 1–48, 2020.

[6] J. Bolte, S. Sabach and M. Teboulle. Proximal alternating linearized minimization for nonconvex and nonsmooth problems. *Mathematical Programming*, 146(1-2): 459–494, 2014.

[7] J.M. Borwein and H. Wolkowicz. Facial reduction for a cone-convex programming problem. *Journal of the Australian Mathematical Society*, 30(3): 369–380, 1980/81.

[8] F. Bunea, C. Giraud, X. Luo, M. Royer and N. Verzelen. Model assisted variable clustering: Minimax-optimal recovery and algorithms. *Annals of Statistics*, 48(1): 111–137, 2020.

[9] Y. Cheung, S. Schurr and H. Wolkowicz. Preprocessing and reduction for degenerate semidefinite programs. *Computational and Analytical Mathematics Springer Proceedings in Mathematics & Statistics*, 50: 251–303, 2013.

[10] A. Chowdhury, P. Drineas, D.P. Woodruff and S. Zhou. Approximation algorithms for sparse principal component analysis. *ICLR*, 2021: 1–25, 2021.

[11] A. d'Aspremont, L. El-Ghaoui, M.I. Jordan and G.R.G. Lanckriet. A direct formulation for sparse PCA using semidefinite programming. *SIAM Review*, 49(3): 434–448, 2007.

[12] A. d'Aspremont, F. Bach and L. El-Ghaoui. Optimal solutions for sparse principal component analysis. *Journal of Machine Learning Research*, 9: 1269–1294, 2008.

[13] A. d'Aspremont, F. Bach and L. El-Ghaoui. Approximation bounds for sparse principal component analysis. *Mathematical Programming*, 148: 89–110, 2014.

[14] D.G. Enki, N.T. Trendafilov and I.T. Jolliffe. A clustering approach to interpretable principal components. *Journal of Applied Statistics*, 40(3): 583–599, 2013.

[15] A. Figueiredo and P. Gomes. Clustering of variables based on watson distribution on hypersphere: A comparison of algorithms. *Communications in Statistics—Simulation and Computation*, 44(10): 2622–2635, 2015.

[16] A. Freitas, E. Macedo and M. Vichi. An empirical comparison of two approaches for CDPCA in high-dimensional data. *Statistical Methods & Applications*, 30(3),11: 1007–1031, 2021.

[17] B. Hajek, Y. Wu and J. Xu. Achieving exact cluster recovery threshold via semidefinite programming: Extensions. *IEEE Transactions on Information Theory*, 62(10): 5918–5937, 2016.

[18] T. Hastie, R. Tibshirani and J.H. Friedman. *The Elements of Statistical Learning: Data Mining, Inference, and Prediction*. Springer, New York, 2009.

[19] D. Henrion, S. Naldi and M. Safey El Din. Exact algorithms for semidefinite programs with degenerate feasible set. *Journal of Symbolic Computation*, 104: 942–959, 2021.

[20] B. Jiang, S. Ma and S. Zhang. Tensor principal component analysis via convex optimization. *Mathematical Programming*, 150(2): 423–457, 2015.

[21] I.T. Jolliffe. *Principal Component Analysis*. Springer, 2002.

[22] I.T. Jolliffe and J. Cadima. Principal component analysis: A review and recent developments. *Mathematical, Physical and Engineering Sciences*, 374, 2016.

[23] I.T. Jolliffe, N.T. Trendafilov and M. Uddin. A modified principal component technique based on the lasso. *Journal of Computational and Graphical Statistics*, 12(3): 531–547, 2003.

[24] J. Kogan, C. Nicholas and M. Teboulle. (eds.). *Grouping Multidimensional Data: Recent Advances in Clustering*. Springer, 2006.

[25] O.I. Kostyukova and T.V. Tchemisova. Optimality criteria without constraint qualification for linear semidefinite problems. *Journal of Mathematical Sciences*, 182(2): 126–143, 2012.

[26] B. Kulis, A.C. Surendran and J.C. Platt. Fast low-rank semidefinite programming for embedding and clustering. *In*: M. Meila and X. Shen (eds.). *Proceedings of the Eleventh International Conference on Artificial Intelligence and Statistics*, volume 2 of *Proceedings of Machine Learning Research*, pp. 235–242. PMLR, 2007.

[27] R. Luss and A. d'Aspremont. Clustering and feature selection using sparse principal component analysis. *Optimization and Engineering*, 11: 145–157, 2010.

[28] S. Ma. Alternating direction method of multipliers for sparse principal component analysis. *Journal of the Operations Research Society of China*, 1: 253–274, 2013.

[29] S. Ma, F. Wang, L. Wei and H. Wolkowicz. Robust principal component analysis using facial reduction. *Optimization and Engineering*, 21: 1195–1219, 2020.

[30] E. Macedo. Testing regularity on linear semidefinite optimization problems. *In*: J.P. Almeida, J.F. Oliveira and A.A. Pinto (eds.). *Operational Research - IO 2013—XVI Congress of APDIO, CIM Series in Mathematical Sciences*, page 213–236. Springer International Publishing, 2015.

[31] E. Macedo. Two-Step-SDP approach to clustering and dimensionality reduction. *Statistics, Optimization, and Information Computing*, 3(3): 294–311, 2015.

[32] E. Macedo and A. Freitas. The alternating least-squares algorithm for CDPCA. *Communications in Computer and Information Science*, 499: 173–191, 2015.

[33] E. Macedo and T. Tchemisova. A generator of nonregular semidefinite programming problems. *In*: A. Vaz, J. Almeida, J. Oliveira and A. Pinto (eds). *Operational Research—APDIO 2017 Springer Proceedings in Mathematics & Statistics*, volume 223, pp. 177–199. Springer, 2018.

[34] A. Majumdar, G. Hall and A.A. Ahmadi. Recent scalability improvements for semidefinite programming with applications in machine learning, control, and robotics. *Annual Review of Control, Robotics, and Autonomous Systems*, 3(1): 331–360, 2020.

[35] F. Martella and M. Vichi. Clustering microarray data using model-based double k-means. *Journal of Applied Statistics*, 2012.

[36] M. McCoy and J.A. Tropp. Two proposals for robust PCA using semidefinite programming. *Electronic Journal of Statistics*, 5: 1123–1160, 2011.

[37] M. Meila. How to tell when a clustering is (approximately) correct using convex relaxations. *In*: S. Bengio, H. Wallach, H. Larochelle, K. Grauman, N. Cesa-Bianchi and R. Garnett (eds.). *Advances in Neural Information Processing Systems*, volume 31, pp. 7407–7418. Curran Associates, Inc., 2018.

[38] A.B. Nieto-Librero, C. Sierra, M.P. Vicente-Galindo, O. Ruíz-Barzola and M.P. Galindo-Villardón. Clustering disjoint hj-biplot: A new tool for identifying pollution patterns in geochemical studies. *Chemosphere*, 176: 389–396, 2017.

[39] M. Olfat and A. Aswani. Convex formulations for fair principal component analysis. In *33rd AAAI Conference on Artificial Intelligence (AAAI-19)*, pp. 663–670, 2019.

[40] B. O'Donoghue, E. Chu, N. Parikh and S. Boyd. Conic optimization via operator splitting and homogeneous self-dual embedding. *Journal of Optimization Theory and Applications*, 169: 1042–1068, 2016.

[41] G. Pataki. Bad semidefinite programs: They all look the same. *SIAM Journal on Optimization*, 27(1): 146–172, 2017.

[42] J. Peng and Y. Wei. Approximating k-means-type clustering via semidefinite programming. *SIAM Journal on Optimization*, 18(1): 186–205, 2007.

[43] A. Pirinen and B. Ames. Exact clustering of weighted graphs via semidefinite programming. *Journal of Machine Learning Research*, 20: 1–34, 2019.

[44] J.A. Ramirez-Figueroa, C. Martin-Barreiro, A. Nieto-Librero, V. Leiva, and M.P. Galindo-Villardón. A new principal component analysis by particle swarm optimization with an environmental application for data science. *Stoch. Environ. Res. Risk Assess*, 35: 1969–1984, 2021.

[45] R. Reris and J. Brooks. Principal component analysis and optimization: A tutorial. In *Proceedings of the 14th INFORMS Comput Soc. Conf.*, pp. 212–225, 2015.

[46] M. Royer. Adaptive clustering through semidefinite programming. *In*: I. Guyon, U.V. Luxburg, S. Bengio, H. Wallach, R. Fergus, S. Vishwanathan and R. Garnett (eds.). *Advances in Neural Information Processing Systems*, volume 30, pp. 1795–1803. Curran Associates, Inc., 2017.

[47] V. Singh, L. Mukherjee, J. Peng and J. Xu. Ensemble clustering using semidefinite programming with applications. *Machine Learning*, 79: 177–200, 2010.

[48] C. Sun, T. Li and V.O.K. Li. Robust and consistent clustering recovery via sdp approaches. In *Proceedings of 2018 IEEE Data Science Workshop (DSW), Lausanne, Switzerland, 4–6 June 2018*, pp. 46–50, 2018.

[49] C. Swamy. Correlation clustering: Maximizing agreements via semidefinite programming. In *Proceedings of the Fifteenth Annual ACM-SIAM Symposium on Discrete Algorithms*, pp. 526–527. SIAM, 2004.

[50] L. Vandenberghe and S. Boyd. Semidefinite programming. *SIAM Review*, 38(1): 49–95, 1996.

[51] M. Vichi and G. Saporta. Clustering and disjoint principal component analysis. *Computational Statistics & Data Analysis*, 53: 3194–3208, 2009.

[52] E. Vigneau, M. Chen and M. Qannari. Clustvarlv: An r package for the clustering of variables around latent variables. *R Journal*, 7: 134, 2015.

[53] H. Waki, M. Nakata and M. Muramatsu. Strange behaviors of interior-point methods for solving semidefinite programming problems in polynomial optimization. *Computational Optimization and Applications*, 53(3): 823–844, 2012.

[54] H. Wei and H. Wolkowicz. Generating and measuring instances of hard semidefinite programs. *Mathematical Programming Ser. A*, 125(1): 31–45, 2010.

[55] J.J. Whang, H. Yangyang, D.F. Gleich and I.S. Dhillon. Non-exhaustive, overlapping clustering. *IEEE Transactions on Patterns Analysis and Machine Intelligence*, 41(11): 2644–2659, 2019.

[56] D.M. Witten and R. Tibshirani. A framework for feature selection in clustering. *Journal of the American Statistical Association*, 105: 713–726, 2010.

[57] H. Wolkowicz and M.F. Anjos. Semidefinite programming for discrete optimization and matrix completion problems. *Discrete Applied Mathematics*, 123: 513–577, 2002.

[58] H. Wolkowicz, R. Saigal and L. Vandenberghe. *Handbook of Semidefinite Programming—Theory, Algorithms, and Applications*. Kluwer Academic Publishers, 2000.

[59] Y. Zhang, A. d'Aspremont and L. El-Ghaoui. Sparse PCA: Convex relaxations, algorithms and applications. *Handbook on Semidefinite, Conic and Polynomial Optimization*, pages 915–940, 2012.

[60] Y. Zhang and L. El-Ghaoui. Large-scale sparse principal component analysis with application to text data. In *Advances in Neural Information Processing Systems (NIPS)*, 2011.

Chapter 5

Operation Research Techniques in Data Mining Focusing on Clustering

Fatma Yerlikaya Özkurt

5.1 Introduction

Recently, data can be collected and stored through various technological sources for different purposes in many fields of study such as biology, medicine, geology, marketing and finance, etc. The analysis of data has a crucial importance in real life activities to obtain valuable information. Traditional methods that are used for processing, analyzing and modeling large databases generally fail. To handle these steps, powerful tools are needed. Data mining methods are extremely preferred and have decisive importance since they are meeting these mentioned steps well.

The main objective of data mining methods is finding meaningful patterns in data and making explicit decisions. The learning methods used in data mining are categorized into two types of learning: *supervised learning* and *unsupervised learning*. In the first type of learning, the aim is to predict the value of an output measure concerning several input measures. Classification and regression tasks

Atilim University.

are considered in this category. On the other hand, in the second type of learning, the outcome measure is not available and the aim is to find the interpretable patterns among given input measures. One of the important tasks in data mining is clustering which is an unsupervised learning method [15].

Clustering is a process that partitions a set of data objects into groups/subsets/clusters based on some similarity criterion. Sometimes, the process of grouping, in other words, labeling data points in different categories may be extremely difficult and time consuming. In such cases, clustering provides a good and useful representation of the data, especially, for large scale data sets [10]. Clustering can be applied to many different types of data in a variety of fields such as computer, medical, earth and social sciences. Besides this, it has also been used a lot in economics, life, and engineering problems for different purposes such as image segmentation, character and object recognition, information retrieval or compression, network analysis, optimal design of mechanical or electrical systems, gene function identification, disease diagnosis, assigning damage states, grouping of customers and firms [16].

Moreover, there are many different approaches/algorithms used to define groups/clusters in different ways. In general, clustering algorithms aim to optimize a specific criterion such as maximizing the intra cluster similarity or minimizing the inter cluster similarity. For that reason, they can be considered optimization problems. The solution of clustering problems is based on the selection of similarity criterion and the objective function, and also using different solution techniques. Selections of different clustering technique or similarity criteria, even for the same clustering techniques with alternative distinct parameters, may produce new clustering results. Therefore, it is not easy to determine the clustering algorithm which is the best in general [28].

On the other part, Operation Research plays an important role in many fields of science and applications. In this chapter, the operations research techniques that include the use of mathematical programming for the formulation of clustering problems are introduced. Reformulation of clustering problems as optimization problems is well-structured and, therefore, the solutions of these problems present competitive and promising results with the help of advanced optimization techniques.

The remaining chapter is organized as follows. In Section 5.2, background on clustering and main clustering methods in data mining are presented. The operations research techniques for the formulation of clustering problems as an optimization problem are given in Section 5.3. In Section 5.4, several operations research applications that use clustering algorithms are provided. Conclusion and outlook are stated in the last section.

5.2 Background on Clustering and Main Clustering Methods in Data Mining

Clustering is an imperative task in data mining and plays an important role in organizing and grouping data appropriately without prior knowledge about the data characteristic. It is a process that aims to identify, specify, and assign data points into different and distinct classes called clusters. In this section, more detailed information about clustering and main clustering methods in data mining will be provided.

5.2.1 Measures of similarity and dissimilarity

Data clustering (or clustering) is one of the most preferred methods in data analysis to create groups/clusters of data points so that data points in one cluster are quite similar and data points in other clusters are considerably distinct. The set of data points that is classified into their respective groups/clusters has to display similar properties based on some criteria or rules. To identify the clusters in a high degree of intra cluster similarity and a high degree of inter cluster dissimilarity, some distance measures named similarity or dissimilarity measures (or in general named as proximity measures) are used [10]. The simple illustration of the intra and inter cluster distance (similarity) for the data set with 60 observations is given in Figure 5.1.

Consider the data set denoted by **D** matrix with N observations, each of which is described by p variables as follows:

$$\mathbf{D} = \begin{pmatrix} x_{11} & x_{12} & \cdots & x_{1p} \\ x_{21} & x_{22} & \cdots & x_{2p} \\ \vdots & \vdots & \ddots & \vdots \\ x_{N1} & x_{N2} & \cdots & x_{Np} \end{pmatrix}. \tag{5.1}$$

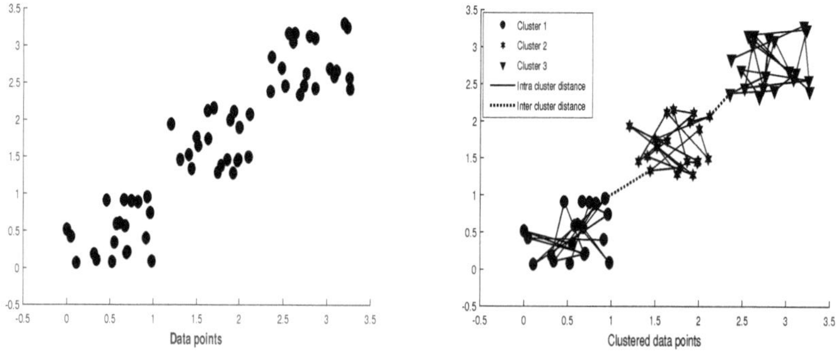

(a) (b)

Figure 5.1: Illustrations of the intra and inter cluster distance for given data set.

Here, $\mathbf{D} = (\mathbf{x}_1, \mathbf{x}_2, ..., \mathbf{x}_N)$, and $\mathbf{x}_i = (x_{i1}, x_{i2}, ..., x_{ip})^T$ is a vector denoting the ith observation and x_{ij} is a data value denoting the jth variable or attribute of $\mathbf{x}_i$ where $i = 1, 2, ..., N$ and $j = 1, 2, ..., p$.

The proximity measures vary according to the data types such as continuous, discrete and mixed that is a combination of the continuous and discrete data types. The continuous data type takes values from an uncountably infinite range set. On the other hand, the discrete data type takes only a finite or a countably infinite number of values. A binary data type which has exactly two values is a special case of discrete data type [28].

For the given data matrix, **D**, the corresponding $(N \times N)$-dimensional proximity matrix can be defined as:

$$\mathbf{P} = \begin{pmatrix} 1 \text{ or } 0 & & & & \\ P(2,1) & 1 \text{ or } 0 & & & \\ P(3,1) & P(3,2) & 1 \text{ or } 0 & & \\ \vdots & \vdots & \vdots & & \\ P(N,1) & P(N,2) & P(N,3) & \cdots & 1 \text{ or } 0 \end{pmatrix}, \tag{5.2}$$

where the $P(m,n)$ element represents the similarity or dissimilarity measure for the mth and nth data points $(m, n = 1, \ldots, N)$. Diagonal elements of this matrix either takes value 1 or 0 based on similarity or dissimilarity concept, respectively. It should be noted that the proximity matrix in Equation 5.2 is constructed as a one-mode form. On the other hand, the proximity for a pair of N dimensional variable $\mathbf{x}_l$ and $\mathbf{x}_k$ is demonstrated by $P(\mathbf{x}_l, \mathbf{x}_k)$ where $l, k = 1, \ldots, N$.

The selection of a proximity measure will affect the clustering result completely, therefore, it is very important to define appropriate proximity measures based on data types. Well-known proximity measures for continuous, discrete and mixed data types are given in Tables 5.1, 5.2 and 5.3, respectively [10, 28].

In Table 5.1, the square matrix inside the formula of the Mahalanobis distance, **S**, is the covariance matrix of the variables. The correlation value used in Pearson correlation, r_{lk}, is defined in Equation 5.3 as follows:

$$r_{lk} = \frac{\sum_{j=1}^{p} (x_{lj} - \bar{x}_l)(x_{kj} - \bar{x}_k)}{\sqrt{\sum_{j=1}^{p} (x_{lj} - \bar{x}_l)^2 \sum_{j-1}^{p} (x_{kj} - \bar{x}_k)^2}}, \tag{5.3}$$

where $\bar{x}_l = (1/p) \sum_{j=1}^{p} x_{lj}$ and $\bar{x}_k = (1/p) \sum_{j=1}^{p} x_{kj}$.

In Table 5.2, the similarity between two binary variables, $P(\mathbf{x}_l, \mathbf{x}_k)$, is obtained using the following four values:

- n_{11} is the number of variables where $\mathbf{x}_l$ is one and $\mathbf{x}_k$ is one.
- n_{00} is the number of variables where $\mathbf{x}_l$ is zero and $\mathbf{x}_k$ is zero.

Table 5.1: Proximity measures for continuous data type.

Measure	**Formula**
Minkowski Distance (L_d Norm)	$P_S(\mathbf{x}_l, \mathbf{x}_k) = \left(\sum_{j=1}^{p} \lvert x_{lj} - x_{kj} \rvert^{1/d}\right)^d$
Euclidean Distance (L_2 Norm)	$P_S(\mathbf{x}_l, \mathbf{x}_k) = \left(\sum_{j=1}^{p} \lvert x_{lj} - x_{kj} \rvert^{1/2}\right)^2$
Manhattan Distance (L_1 Norm)	$P_S(\mathbf{x}_l, \mathbf{x}_k) = \sum_{j=1}^{p} \lvert x_{lj} - x_{kj} \rvert$
Maximum Distance (L_∞ Norm)	$P_S(\mathbf{x}_l, \mathbf{x}_k) = \max_{1 \le j \le p} \lvert x_{lj} - x_{kj} \rvert$
Mahalanobis Distance	$P_S(\mathbf{x}_l, \mathbf{x}_k) = (\mathbf{x}_l - \mathbf{x}_k)^T \mathbf{S}^{-1} (\mathbf{x}_l - \mathbf{x}_k)$
Pearson Correlation	$P_S(\mathbf{x}_l, \mathbf{x}_k) = (1 - r_{lk})/2$

Table 5.2: Proximity measures for discrete data type.

Measure	**Formula**
Similarity Coefficients	$P_S(\mathbf{x}_l, \mathbf{x}_k) = \frac{n_{11}+n_{00}}{n_{11}+n_{00}+w(n_{10}+n_{01})}$
Hamming Distance	$P_S(\mathbf{x}_l, \mathbf{x}_k) = \frac{n_{11}}{n_{11}+w(n_{10}+n_{01})}$
Simple Matching Coefficient	$P_S(\mathbf{x}_l, \mathbf{x}_k) = \frac{1}{p} \sum_{j=1}^{p} S_{lkj}$

Table 5.3: Proximity measures for mixed data type.

Measure	**Formula**
Similarity Measure	$P_S(\mathbf{x}_l, \mathbf{x}_k) = \frac{\sum_{j=1}^{p} S_{lkj} \delta_{lkj}}{\sum_{j=1}^{p} \delta_{lkj}}$

- n_{10} is the number of variables where $\mathbf{x}_l$ is one and $\mathbf{x}_k$ is zero.
- n_{01} is the number of variables where $\mathbf{x}_l$ is zero and $\mathbf{x}_k$ is one.

Moreover, the weight, w, can take 1, 2 or 0.5 based on unmatched pairs' importance. In simple matching coefficient, the similarity that is used for discrete variables with more than two values, S_{lkj}, is defined as follows:

$$S_{lkj} = \begin{cases} 0 & \text{if } \mathbf{x}_l \text{ and } \mathbf{x}_k \text{ do not match in the } j\text{th variable,} \\ 1 & \text{if } \mathbf{x}_l \text{ and } \mathbf{x}_k \text{ match in the } j\text{th variable.} \end{cases} \tag{5.4}$$

In Table 5.3, the δ_{lkj} is a coefficient based on missing observation and the similarity components both for discrete (including binary case) and continuous variables can be obtained as follows:

$$S_{lkj} = \begin{cases} 1 & \text{if } \mathbf{x}_{lj} = \mathbf{x}_{kj}, \\ 0 & \text{if } \mathbf{x}_{lj} \neq \mathbf{x}_{kj}, \end{cases} \tag{5.5}$$

and

$$S_{lkj} = 1 - \frac{|x_{lj} - x_{kj}|}{\max x_j - \min x_j}, \tag{5.6}$$

respectively. Correspondingly, the dissimilarity measure for all cases (discrete, continuous and mixed type variables) can be obtained by simply using the formula $P_D(\mathbf{x}_l, \mathbf{x}_k) = 1 - P_S(\mathbf{x}_l, \mathbf{x}_k)$ [10, 28].

5.2.2 Clustering methods in data mining

The actual cluster structures and the number of clusters in a given data set may not be known in advance because of the nature of the data sets. Using different similarity measures based on the characteristics of the data set can end up obtaining different clustering results. There are several clustering methods in the literature to deal with these issues. In this subsection, clustering methods in data mining are classified into four categories and are briefly described [15].

5.2.2.1 Hierarchical clustering

Hierarchical clustering aims to construct clusters that have an ordering from bottom to top like a tree structure. As a result, it produces the hierarchical relation between the created clusters. There are two kinds of hierarchical clustering named divisive and agglomerative [18]. Divisive clustering is splitting the single all inclusive cluster into two until having only clusters with one data point. Whereas agglomerative clustering (bottom-up approaches) is starting from the single data point as an individual cluster and merging clusters at each iteration until getting a single all inclusive cluster. As a proximity measure, the linkage metrics such as single, complete, and average linkage can be used to represent a cluster.

The number of clusters is not needed to be predefined in hierarchical clustering. Indeed, this is the main advantage of hierarchical clustering. On the other hand, being computationally expensive a disadvantage since it constructs the proximity matrix between each data point and also has no flexibility of applying an adjustment for the iterations to merge or split data points [18].

5.2.2.2 *Partitioning clustering*

Partitioning clustering is based on clustering of N unlabeled data points to K clusters in which each cluster contains at least one data point. The purpose of the partitioning clustering is to minimize the distances of data points in a cluster and to maximize the distances between the separated clusters. In partitioning clustering, after defining the number of clusters (K), the next step is to assign K random initial centers. The cluster centers are updated based on the data points assigned to a given cluster. This procedure repeatedly continues until the assigned cluster points of a sample can not be updated [10].

The commonly used partitioning clustering methods are K-means and partitioning around medoids (PAM). In both methods, each data point is assigned to only one cluster and the clustering solution differs depending on the initial cluster centers [27]. The advantage of partitioning clustering to hierarchical clustering is to have rapid computational evaluation.

5.2.2.3 *Density based clustering*

Density based clustering is a clustering method that identifies the arbitrarily shaped clusters in data according to the idea of a cluster is being a region with high density and separated from the other such clusters by regions of low density [19]. Although these types of clustering algorithms have high complexity, they can easily identify outliers in the data set. Moreover, they can handle noise and detect the clusters automatically since they can scan the data well.

5.2.2.4 *Grid based clustering*

The grid based clustering is different from the other clustering methods in a such way that it is based on the value space around the data points rather than data points. The general procedure of the grid based clustering algorithm is starting with creating a grid structure. After partitioning the data space into a number of regions like square, rectangle or other such types of forms, the calculation of this region density for each of the regions is fulfilled. Finally, these regions are sorted with respect to their densities and the cluster centers are defined [23].

Indeed, the evaluation of the clustering results is not straight forward due to unknown label information. It should be noted that a different clustering method, even for the same method with different alternative parameters, can produce different clustering results. Figure 5.2 illustrates the subjectivity of the three main clustering methods (hierarchical clustering, density based clustering, partitioning). While the first and second clustering methods ((a) and (b)) lead to two clusters with different partitioning, the last one ((c)) leads to three clusters [24].

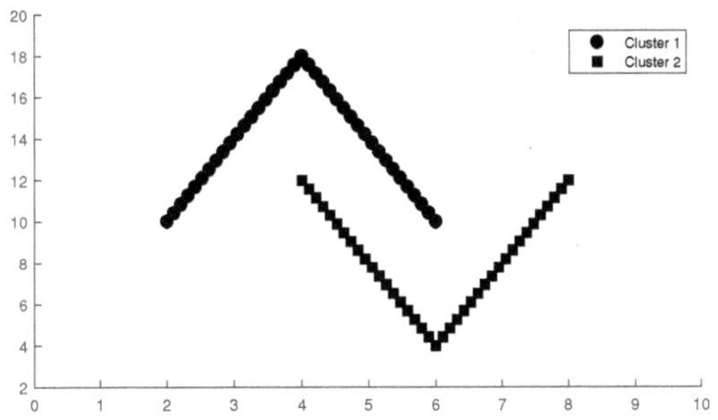

(a) Hierarchical Clustering

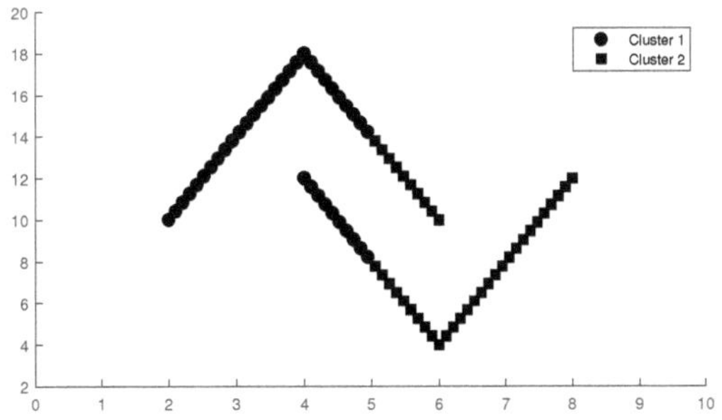

(b) Density Based Clustering

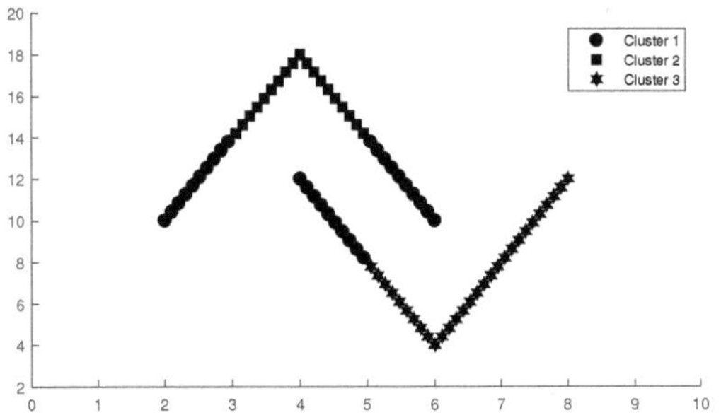

(c) Partitioning Clustering

Figure 5.2: Illustration of subjectivity of three main clustering methods: hierarchical clustering, density based clustering, partitioning clustering.

5.3 Formulation of Clustering Problems as an Optimization Problem

The clustering algorithms may be seen as optimization problems since they try to optimize a specific criterion that improves the clustering results. In the past three decades, there has been an increasing in the development of clustering techniques based on modeling and constraint optimization in a wide range of application areas. Both continuous optimization and integer optimization methods have been widely used and have significantly improved the clustering methods mentioned in Section 5.2 [1]. The general procedure for handling clustering problem as an optimization problem can be summarized as follows:

1. Data preparation (including data standardization and/or transformation);
2. Modelling (construction of objective function and constraints);
3. Optimization (using appropriate optimization techniques to solve the problem);
4. Validation (evaluation of the results based on the unseen data set).

In clustering problems, construction of objective function and constraints is an important issue. The clustering problem should be stated in a such a way that existence clusters have to be well-separated and dissimilar data points aim to be grouped into different clusters. Consequently, the distance between clusters should be maximized. Moreover, similar data points have to be placed into the same cluster and should be close to each other. This is called minimization of the compactness measures. Thus, the formed clusters are associated with the type of the proximity measure.

Consider that for given data points, $\mathbf{D} = (\mathbf{x}_1, \mathbf{x}_2, ..., \mathbf{x}_N)$, the clustering algorithm intends to group them into K clusters: $\{C_1, C_2, \ldots, C_K\}$ to maximize or minimize a determined proximity function J. Suppose that the proximity is defined as the sum of squared error. Then, the minimization problem of the objective function, J, is as follows:

$$\begin{aligned} J &= \sum_{c=1}^{K} \sum_{i=1}^{N} \beta_{ci} \left\| \mathbf{x}_i - \mathbf{m}_c \right\| \\ &= \sum_{c=1}^{K} \sum_{i=1}^{N} \beta_{ci} \left(\mathbf{x}_i - \mathbf{m}_c \right)^T \left(\mathbf{x}_i - \mathbf{m}_c \right). \end{aligned} \tag{5.7}$$

Here, the cluster prototype matrix is $\mathbf{M} = [\mathbf{m}_1, \mathbf{m}_2, \ldots, \mathbf{m}_K]$ and the sample mean for the cth cluster with N data points is $\mathbf{m}_c = \frac{1}{N} \sum_{i=1}^{N} \mathbf{x}_i$. Moreover,

$$\beta_{ci} = \begin{cases} 1 & \text{if } \mathbf{x}_i \text{ is the element of cluster } C_c, \\ 0 & \text{otherwise.} \end{cases}$$

where $\sum_{c=1}^{K} \beta_{ci} = 1$ for all i.

The generalized form of the objective function can be presented as follows:

$$J = \sum_{c=1}^{K} \psi_1(i) = \sum_{c=1}^{K} \frac{1}{N} \sum_{\mathbf{x}_l \in C_c} \sum_{\mathbf{x}_k \in C_c} P(\mathbf{x}_l, \mathbf{x}_k). \tag{5.8}$$

In Equation 5.8, $P(\mathbf{x}_l, \mathbf{x}_k)$ defines the proximity between two data points $\mathbf{x}_l$ and $\mathbf{x}_k$ in C_c cluster and it simply measures the intra cluster similarity of this cluster (see Subsection 5.2.1).

It is also possible to extend the clustering problem by using a different mathematical formulation for the objective function that leads to a different optimization problem. In the following, the most important and preferred ones are introduced [28]:

1. The objective function can measure the maximum distance between two different data points $\mathbf{x}_l$ and $\mathbf{x}_k$ inside C_cth cluster:

$$\psi_2(i) = \max_{\mathbf{x}_l \in C_c \ \mathbf{x}_k \in C_c \ l \neq k} \left(P_D(\mathbf{x}_l, \mathbf{x}_k) \right). \tag{5.9}$$

2. The objective function can measure the minimum sum of the distances between all pairs of data points of C_c cluster:

$$\psi_3(i) = \min_{\mathbf{x}_l \in C_c} \left(\sum_{\mathbf{x}_k \in C_c} P_S(\mathbf{x}_l, \mathbf{x}_k) \right). \tag{5.10}$$

3. The objective function can measure the minimum distance for a data point $\mathbf{x}_l$ inside C_cth cluster and the maximum distance between the data points $\mathbf{x}_l$ and $\mathbf{x}_k$ that is not placed in C_cth cluster:

$$\psi_4(i) = \min_{\mathbf{x}_l \in C_c} \left(\max_{\mathbf{x}_k \notin C_c} P(\mathbf{x}_l, \mathbf{x}_k) \right). \tag{5.11}$$

4. The objective function can measure the sum of the distances for all data points inside C_cth cluster and the other points out side of C_c cluster:

$$\psi_5(i) = \sum_{\mathbf{x}_l \in C_c} \sum_{\mathbf{x}_k \notin C_c} P_S(\mathbf{x}_l, \mathbf{x}_k). \tag{5.12}$$

5. The objective function can measure the minimum distance between a point $\mathbf{x}_l$ of C_c cluster and all other points placed out of C_c cluster:

$$\psi_6(i) = \min_{\mathbf{x}_l \in C_c \ \mathbf{x}_k \notin C_c} P_S(\mathbf{x}_l, \mathbf{x}_k). \tag{5.13}$$

The clustering problem defined as an optimization problem may have to optimize one or more constraint functions that increase the quality of the clustering results. These constraints are useful not only to reduce the clustering algorithm complexity but also to find best cluster structures that fit given data points. Therefore, in this chapter, the construction of the optimization problems by using constraint programming is also introduced. In the following, some of the most important constraints are given.

- **Connection-Required Constraint:** Two data points $\mathbf{x}_l$ and $\mathbf{x}_k$ $(l, k = 1, 2, \ldots, N)$ must belong to the same cluster $\mathbf{x}_l, \mathbf{x}_k \in C_c$.

- **Connection-Not Required Constraint:** Two data points $\mathbf{x}_l$ and $\mathbf{x}_k$ must belong to the different cluster $\mathbf{x}_l \in C_l, \mathbf{x}_k \in C_k$.
- **Cluster Size Constraint:** The size of the cluster should be determined with respect to a minimum or maximum numbers of data points such that $|C_c| \geq \kappa$ or $|C_c| \leq \kappa$.
- **Cluster Dimension Constraint:** For each pairwise data points $\mathbf{x}_l$ and $\mathbf{x}_k$ in the same cluster, the dimension (d) of the each cluster should be at most ξ so that $d(\mathbf{x}_l, \mathbf{x}_k) \leq \xi$.
- **Margin Constraint:** For any two different clusters (C_l and C_k), the clusters margin should be higher than a specific threshold value ς so that $\mathbf{x}_l \in C_l$, $\mathbf{x}_k \in C_k$ and $d(\mathbf{x}_l, \mathbf{x}_k) \geq \varsigma$.

Here, C_c $(c = 1, 2, \ldots, K)$ is the cth cluster [13].

As an example, firstly, the following optimization problem which is the minimization of the sum of squared error (SSE) with respect to some constraints is considered. For all data points: $\mathbf{D} = (\mathbf{x}_1, \mathbf{x}_2, ..., \mathbf{x}_N)$ and K clusters: $\mathbf{C} = \{C_1, C_2, \ldots, C_K\}$, the squared distance between ith observation and mth medoid in cth cluster is defined so that ith data point can only belong to one cluster ($C_l \cap C_k = \emptyset$) and each cluster can have only one medoid [13]. The corresponding constraint optimization problem formulation is for one of the partitioning clustering algorithm which is a K-medoid clustering and given as follows:

$$\min_{\mathbf{C}} \ \mathrm{SSE}(\mathbf{C}) \tag{5.14}$$

subject to

$$\begin{aligned}
& a_{i,C_c} = 1 \ \text{ if } \ \mathbf{x}_i \ \text{ is assigned to cluster } \ C_c, \\
& a_{i,C_c} = 0 \ \text{ if } \ \mathbf{x}_i \ \text{ is not assigned to cluster } \ C_c, \\
& m_{i,C_c} = 1 \ \text{ if } \ \mathbf{x}_i \ \text{ is the medoid of cluster } \ C_c, \\
& m_{i,C_c} = 0 \ \text{ if } \ \mathbf{x}_i \ \text{ is not the medoid of cluster } \ C_c, \\
& SSE(C_c) = \sum_{\mathbf{x}_i \in \mathbf{D}} a_{i,C_c} \sum_{\mathbf{x}_i \in \mathbf{D} \ \& \ \mathbf{x}_k \neq \mathbf{x}_i} m_{i,C_c} \, d^2(\mathbf{x}_i, \mathbf{x}_k), \\
& \sum_{C_c \in \mathbf{C}} a_{i,C_c} = 1 \ \text{ for each data points}, \\
& \sum_{\mathbf{x}_i \in \mathbf{D}} m_{i,C_c} = 1 \ \text{ for each clusters}, \\
& \left| \bigcup_{C_c \in \mathbf{C}} C_c \right| = |\mathbf{D}|, \\
& |\mathbf{C}| = K.
\end{aligned}$$

The following constraint optimization problem formulation is constructed based on significant advances in mixed-integer optimization methods to the problems of density based clustering. The optimization model searches for the maximum number of different labels [13].

$$\max \left(\sum_{l_i \in \mathbf{L}} \min \left(1, \sum_{\mathbf{x}_i \in \mathbf{D}} t_{i,l_i} \right) \right) \tag{5.15}$$

subject to

$a_{l,k} = 1$ if distance between any two points in **D**: $d(\mathbf{x}_l, \mathbf{x}_k) \leq \varepsilon$,

$a_{l,k} = 0$ if distance between any two points in **D**: $d(\mathbf{x}_l, \mathbf{x}_k) > \varepsilon$,

$t_{l,l_i} = 1$ if data point $\mathbf{x}_l \in \mathbf{D}$ has label $l_i \in \mathbf{L}$,

$t_{l,l_i} = 0$ if data point $\mathbf{x}_l \in \mathbf{D}$ has not label $l_i \in \mathbf{L}$,

$r_l = 1$ if data point $\mathbf{x}_l \in \mathbf{D}$ is a core point means that $\sum_{\mathbf{x}_k \in \mathbf{D}} a_{l,k} \geq minr$,

$r_l = 0$ if data point $\mathbf{x}_l \in \mathbf{D}$ is not a core point,

$\sum_{l_i \in \mathbf{L}} t_{l,l_i} = 1$ for each data points,

$r_l = 1$ and $r_k = 1$ and $a_{l,k} = 1$ and $t_{l,l_l} = 1 \Rightarrow t_{k,l_l} = 1$,

$r_l = 0 \Rightarrow t_{l,l_l} = 1$.

Here, $\mathbf{L} = \{l_1, ..., l_N, l_{N+1}\}$ is a vector that includes an ordered set of labels and

$$l_i = \min \left\{ \{ l_i \in \mathbf{L} \setminus \{l_{N+1}\} \mid a_{l,k} = 1 \text{ and } r_l = 1 \text{ and } t_{l,l_l} = 1 \} \cup \{l_{N+1}\} \right\}.$$

In Equation 5.15, the 8th constraint is intended to obtain a data point that has just one label, and all connected core points should have the same label. The last constraint is aimed to attend a data point to a different label if there is not a core point connected to the data point. On the other hand, if the data point does not connect to any core point, then this data point can have an additional label l_{N+1} [13].

To solve the optimization problems introduced in this section, there are many different types of search algorithms used for finding the global or approximate global optimum solution. For example, genetic algorithms are suitable for solving complex clustering problems to reach the optimal solution. They have been successfully applied to hierarchical and partitioning clustering to improve clustering results [6, 12, 22]. Simulated and deterministic annealing are sequential and global search techniques that used to solve clustering problems by assigning a certain probability to a cluster [2, 25]. The method of Lagrange optimization is another way of solving the minimum error criterion function given in Equations 5.8 to 5.13. There are also many stochastic and fuzzy type solution algorithms used for the solution of clustering problems.

5.4 Operations Research Applications of Clustering Algorithms

Data can be collected due to several reasons and for different purposes. After obtaining data from different sources, analyzing and clustering data become more important, and to conduct these steps, advanced mathematical and statistical techniques are needed. One of the most important approaches in data analysis is the use of operations research techniques.

Clustering algorithms that use operations research techniques can be applied to distinct types of data in a variety of fields [8]. Some applications of clustering concerning different fields are listed below.

- ***Engineering*:** There are many different applications of clustering in a branch of engineering such as industrial engineering, aerospace engineering, chemical engineering, civil engineering, electrical engineering, mechanical engineering. The use of clustering in these fields aims to find the best matches in data sets. For example, high dimensional engineering designs can be formulated as clustering problems to obtain groups of similar parts or elements, each of which is homogeneous according to specific design features. In this way, duplication of design efforts will be reduced [4, 7, 11].

- ***Computer Sciences*:** Advances in hardware and software technologies enable us to do more and more applications of clustering for huge data sets available in computer sciences. Especially, document clustering in web mining and text mining plays an important role in disclosing useful information from documents and organizing query results obtained from a search engine. Other implementations of clustering are conducted for object and character recognition, partitioning an image into regions (image segmentation), organizing documents or databases to browse and explore easily by the users (information retrieval) [3, 5, 14, 16].

- ***Life and Medical Sciences*:** The areas such as genetics, biology, microbiology, paleontology, and psychiatry consist of important applications of clustering. Clustering algorithms are used and will continue to be used in these areas to reveal real patterns in hidden structures of the corresponding data sets. Well known applications of clustering algorithms are protein and gene expression, taxonomy definition, disease diagnosis and treatment. Clustering algorithms are preferred, for example, in gene expression analysis for revealing bio molecular clustering of tumor samples; in disease diagnosis for grouping of the internal tissues and for defining the location of pathology in the living organism [17, 20, 21].

- ***Astronomy and Earth Sciences*:** In earth sciences such as geography and geology, clustering algorithms are used to investigate land formations, group regions and cities, and to understand complex systems of rivers or mountains. On the other hand, in astronomy, in the applications of remote sensing, clustering is preferred to define and classify elements of the universe such as stars and planets [9, 16].

- ***Social Sciences*:** Recently, interesting applications of clustering can be found in the field of social sciences such as sociology, psychology, archaeology, anthropology, and education. For example, behavior pattern analysis based on clustering is used in criminal psychology. On the other hand, clustering a population into a homogeneous community and identifying and establishing the relationship between different cultures is important clustering analysis for social networks. Recently, the use of clustering algorithms has gained importance for archaeological findings and artifacts [10, 28].

- ***Economics and Finance*:** The most up-to-date and popular applications based on clustering are in marketing and business. A few examples of clustering applications are classifying customer characteristics, grouping of firms, analyzing and finding out similar stock price trend [10, 26].

5.5 Conclusion and Outlook

Clustering is one of the useful and important tasks in data mining. Clustering aims is to group similar data points into one cluster while partitioning dissimilar data points into different clusters. In this chapter, main clustering methods and measures of similarity and dissimilarity have been revisited.

The clustering problem can be formulated in terms of a real function of several real variables, and a set of arguments that give an optimal clustering. Such a problem may arise from parameter estimation in data mining methods, and clustering as a special case of this is considered in this chapter. Since the most of the clustering algorithms can be considered as optimization problems, operations research techniques are preferred and widely used in the literature. As known, operation research plays an important role in many fields of science and it has wide range of uses in different real world applications. In this book chapter, clustering algorithms that use operations research techniques are discussed. Finally, several operations research applications that use clustering algorithms are provided in this chapter.

References

[1] T.S. Arthanari. Mathematical programming in statistics. Technical report, 1981.

[2] D.E. Brown and C.L. Huntley. A practical application of simulated annealing to clustering. *Pattern Recognition*, 25(4): 401–412, 1992.

[3] D. Cai, X. He and J. Han. Document clustering using locality preserving indexing. *IEEE Transactions on Knowledge and Data Engineering*, 17(12): 1624–1637, 2005.

[4] T.P. Caudell, S.D.G. Smith, R. Escobedo and M. Anderson. Nirs: Large scale art-1 neural architectures for engineering design retrieval. *Neural Networks*, 7(9): 1339–1350, 1994.

[5] M.E. Celebi, Y.A. Aslandogan and P.R. Bergstresser. Mining biomedical images with density-based clustering. In *International Conference on Information Technology: Coding and Computing (ITCC'05)-volume II*, volume 1, pp. 163–168. IEEE, 2005.

[6] C.-H. Cheng, W.-K. Lee and K.-F. Wong. A genetic algorithm-based clustering approach for database partitioning. *IEEE Transactions on Systems, Man, and Cybernetics, Part C (Applications and Reviews)*, 32(3): 215–230, 2002.

[7] P.J. Componation and J. Byrd. Utilizing cluster analysis to structure concurrent engineering teams. *IEEE Transactions on Engineering Management*, 47(2): 269–280, 2000.

[8] B. Everitt, S. Landau, M. Leese and D. Stahl. *Cluster Analysis*. John Wiley and Sons, 2011.

[9] B.S. Everitt, S. Landau and M. Leese. Cluster analysis arnold. *A Member of the Hodder Headline Group, London*, pp. 429–438, 2001.

[10] G. Gan, C. Ma and J. Wu. Data clustering: Theory, algorithms, and applications. *SIAM*, 2020.

[11] A.N. Gorban, A.A. Pitenko, A.Y. Zinovyev and D.C. Wunsch. Visualization of any data with elastic map method. *Proceeding of Artificial Neural Networks in Engineering, St. Louis*, 2001.

[12] W.A. Greene. Unsupervised hierarchical clustering via a genetic algorithm. In *The 2003 Congress on Evolutionary Computation, 2003. CEC'03.*, volume 2, pp. 998–1005. IEEE, 2003.

[13] V. Grossi, A. Monreale, M. Nanni, D. Pedreschi and F. Turini. Clustering formulation using constraint optimization. In *SEFM 2015 Collocated Workshops*, pp. 93–107. Springer, 2015.

[14] K.M. Hammouda and M.S. Kamel. Efficient phrase-based document indexing for web document clustering. *IEEE Transactions on Knowledge and Data Engineering*, 16(10): 1279–1296, 2004.

[15] J. Han, M. Kamber and J. Pei. Data mining concepts and techniques third edition. *The Morgan Kaufmann Series in Data Management Systems*, 5(4): 83–124, 2011.

[16] A.K. Jain, M.N. Murty and P.J. Flynn. Data clustering: A review. *ACM Computing Surveys (CSUR)*, 31(3): 264–323, 1999.

[17] D. Jiang, C. Tang and A. Zhang. Cluster analysis for gene expression data: A survey. *IEEE Transactions on Knowledge and Data Engineering*, 16(11): 1370–1386, 2004.

[18] L. Kaufman and P.J. Rousseeuw. *Finding Groups in Data: An Introduction to Cluster Analysis*. John Wiley & Sons, 2009.

[19] H.-P. Kriegel, P. Kröger, J. Sander and A. Zimek. Density-based clustering. *Wiley Interdisciplinary Reviews: Data Mining and Knowledge Discovery*, 1(3): 231–240, 2011.

[20] A.W.-C. Liew, H. Yan and M. Yang. Pattern recognition techniques for the emerging field of bioinformatics: A review. *Pattern Recognition*, 38(11): 2055–2073, 2005.

[21] S.C. Madeira and A.L. Oliveira. Biclustering algorithms for biological data analysis: A survey. *IEEE/ACM Transactions on Computational Biology and Bioinformatics*, 1(1): 24–45, 2004.

[22] U. Maulik and S. Bandyopadhyay. Genetic algorithm-based clustering technique. *Pattern Recognition*, 33(9): 1455–1465, 2000.

[23] A.H. Pilevar and M. Sukumar. Gchl: A grid-clustering algorithm for high-dimensional very large spatial data bases. *Pattern Recognition Letters*, 26(7): 999–1010, 2005.

[24] T. Ronan, Z. Qi and K.M. Naegle. Avoiding common pitfalls when clustering biological data. *Science Signaling*, 9(432): re6, 2016.

[25] S.Z. Selim and K.1. Alsultan. A simulated annealing algorithm for the clustering problem. *Pattern Recognition*, 24(10): 1003–1008, 1991.

[26] J. Shadbolt. *Neural Networks and the Financial Markets: Bpredicting, Combining, and Portfolio Optimisation.* Springer Science & Business Media, 2002.

[27] S. Theodoridis and K. Koutroumbas. Linear classifiers. *Pattern Recognition. 3rd ed. San Diego, Calif: Academic Press*, pp. 69–78, 2006.

[28] R. Xu and D. Wunsch. *Clustering*. John Wiley & Sons, 2008.

Chapter 6

Data Mining Approaches to Meteorological Data: A Review of NINLIL Climate Research Group Studies

İnci Batmaz

6.1 Introduction

Global warming is one of the most important concerns that humans face in the 21st century. According to the International Panel on Climate Change (IPCC)'s report [34], the earth's global average surface temperature has increased by about 0.85 degrees centigrade over the last century. In addition, the types and variability of the other climatic elements have been changing as well. If climate change is not detected and treated carefully, it will lead to disasters such as floods and droughts. Therefore, climate change has engaged the attention of the researchers and has became the focus of recent studies. Data Science provides these researchers new approaches such as Data Mining (DM) to let them analyze meteorological (climate) data thoroughly manner. Knowledge discovery in

NINLIL Climate Research Group, Middle East Technical University, Department of Statistics, Turkey (http://stat.metu.edu.tr/en/ninlil-research-group).

databases (KDD) is an iterative process for discovering interesting patterns in large datasets. DM is an important step in the KDD process where data analysis and discovery algorithms are implemented [11].

Turkey, situated in the Mediterranean Basin, is one of the regions which will face the negative impacts of climate change such as variability in hydrology, ecosystem and agriculture related to extreme events [36]. Here, NINLIL is a Climate Research Group formed in 2006 for the purpose of investigating the effects of globally experienced climate change on Turkey. It is an interdisciplinary research group consisting of data scientists from various disciplines including Statistics, Geography, Industrial and Computer Engineering, and Applied Mathematics. During the last one and a half decade, the group has carried out research and developments on climate change by making use of the KDD process, and DM, particularly, to search for answers to the following questions:

Has the climate been changing in Turkey? If so;

- Identify the boundaries of the current climate regions.
- Determine how many seasons Turkey goes through within a year and define them.
- Develop early warning (surveillance) systems for the regions under the risk of floods or droughts.

To accomplish this purpose, the meteorological data consists of 50 climatological variables recorded at 277 stations during 60 years obtained from the Turkish State of Meteorological Service (TSMS) along with the 'metadata' containing stations' information is utilized.

Climate data obtained from the stations have some special features peculiar to itself; it contains several variables measured with different metrics, and also has many incomplete (missing) and inconsistent values. Therefore, rigorous preprocessing of the data is essential before analyzing it. Useful and powerful tools are available thanks to the KDD process. After the preprocessing, the data can be explored safely for an 'unexpected and/or extraordinary' structure that may hide in it.

In this chapter, the research results of the NINLIL group are reviewed to exemplify the KDD process and DM approaches that can be applied to the meteorological data. In Section 6.2, KDD and DM are briefly introduced with their functions and methods. Turkish meteorological data utilized is described and preprocessed in Section 6.3. In Section 6.4, descriptive mining, clustering and modeling applications that respond to the above questions posed to the Turkish climate data are given. The experiences gained and future studies are presented in the last section.

6.2 Knowledge Discovery in Databases and Data Mining

KDD is an iterative search process of exploring unusual patterns in data [23]. The basic steps of this process are data preparation and preprocessing, DM, evaluation/interpretation, and implementation. Based on the needs of the dataset, these steps can be visited as many times as required. Note here that DM is an important step in the KDD process where data is analyzed by using descriptive and predictive DM functions. These functions include association, clustering, trend analysis, similarity analysis, pattern detection, periodicity analysis and so on. The steps are described briefly below.

6.2.1 *Data preparation and preprocessing*

DM functions generally process the data stored systematically in a structured way in flat files, relational and/or transactional databases, data warehouses, and data marts [32]. In certain applications, data is stored in different repositories. So that first of all, the sources are accessed and integrated, and then, put into a table format where rows and columns represent cases (i.e., records/observations) and variables, respectively [22]. When big data is formed as a result of this step, a sample of it is processed instead, or alternatively, big data processing mediums are utilized where available [1].

Sometimes, more complex unstructured data like images, multimedia, or hypertext data stored in object-oriented databases (e.g., spatio-temporal databases in our case) have to be processed. For example, meteorological data obtained from satellites are usually stored in such databases. In this chapter, only structured data gathered from meteorological stations spread out in Turkey is studied.

On the other hand, data preprocessing is concerned about treating dirty, incomplete and inconsistent real-life data to improve its quality [1, 23]. We should keep in mind that reliable results can only be obtained using quality data according to the 'garbage in garbage out' principle. Afterward, data transformation and/or reduction are applied to improve the efficiency of the DM functions. Main data preprocessing functions can be listed as follows [23]:

- Data cleaning deals with imputing missing values, detecting and treating outliers, and eliminating redundant data.
- Data transformation involves putting data into appropriate forms when necessary. Applying mathematical functions, normalization, smoothing, aggregation, generalizations, and discretization are examples for such transformations.
- Data reduction involves in either reducing the size of the data by using statistical models, histograms, clusters and samples, or reducing the num-

ber of variables contained by using dimension reduction techniques, also known as feature selection.

6.2.2 Data mining

DM functions can be grouped into two: descriptive mining and predictive mining [17, 25]. While descriptive functions detect patterns, predictive ones enable to construct models between variables for predicting future values.

- Descriptive Mining Functions:
  - *Summarization* helps to describe data [22, 28]. Descriptive statistics, graphical displays, dependency and association measures are some such examples.
  - *Clustering* puts similar objects into a group by utilizing the distance measures like Euclidean and Manhattan distance [28]. Clustering analysis can be used for both describing and preprocessing data. Some examples include partitioning methods (e.g., k-means, partitioning around medoids (PAM), fuzzy k-means), hierarchical methods (e.g., agglomerative and divisive approach which includes single, average and complete linkage), density-based, grid-based, model-based methods, and the ones merging individual model abilities like ensemble and consensus methods.
  - *Association* tries to determine the items occurring altogether. It first finds frequent item sets, and then, generates interesting rules based on them measured by support (i.e., frequency of occurrence), confidence (i.e., strength of the rule) and lift (i.e., relation of confidence to support) [24].
  - *Apriori* is a well-known association algorithm extensively used in the literature.
- Predictive Mining Functions:
  - *Classification* involves constructing a model for the qualitative type of response and categorizing a new dataset, accordingly. Some examples are statistical-based methods (e.g., Generalized Linear Models (GLM), Bayesian classifiers), Decision-Tree (DT)-based ones (e.g., ID3, C4.5, CAID, CART), Artificial Neural Networks (ANN)-based methods (e.g., Radial-Basis Function (RBF), Competitive ANN (SOM)), Bayesian NN and the others (e.g., K-Nearest-Neighbors (KNN), Rough Set Theory (RST), Support Vector Machines (SVM), the combination of multiple classifiers like consensus clustering, boosting and bagging which improve the classifiers' accuracy).

- *Prediction* involves forecasting a continuous type of response variable. Some of the well-known methods are Multiple Linear Regression (MLR), Time Series Analysis (TSA), Classification and Regression Trees (CART), Multivariate Adaptive Regression Splines (MARS), ANN, and DT-based ones.

6.2.3 Evaluation, interpretation and implementation

To obtain reliable results, the knowledge extracted by applying proper DM functions on the data should be rigorously evaluated and compared for some performance criteria such as accuracy, precision, stability, robustness and computation time [17]. Nevertheless, objective evaluation of the results may not always come up with a sound outcome. In such a case, domain expert judgement can be very useful for a realistic interpretation of the results. Finally, implementation is the last stage that distinguishes DM applications from other types of data analysis. In addition to correct interpretation, special dissemination and implementation approaches developed by the domain experts to highlight the DM results, are essential. As a result, from the Data Science's perspective, it can be said that domain experts are indispensable research partners for the success of the DM applications.

6.3 Climate Data Preparation and Preprocessing

TSMS have recorded various variables since the mid-1920s by manually until 2003, both manually and automatically between 2003 and 2006; then, automatically only. However, data collected by the 1950s are not trustable because the quality checks have not been done yet [35]. The locations of stations are selected to be outside of the residence area. Nevertheless, due to unplanned urbanization in Turkey, today some of the stations (27 stations out of 277; i.e., 26%) reside in the urban area. Besides, while their density changes between 2.59 and 5.08, on average, 3.4 stations fall per 10.000 km^2 in Turkey. Moreover, a significant increase in the number of stations is observed at the beginning of the 1960s, and it stays mostly stable afterwards. It is good to know that as the number of automatic stations increases, they will be spread out in the rural area, and thus, enabling microclimatic features of our country to be recorded.

Turkish meteorological data obtained from TSMS consists of monthly observations of 34 climatological variables (e.g., precipitation, temperature, humidity) and daily observations of 16 climatological variables recorded at 277 stations, which are well distributed all over the country, from 1950 to 2006 into an electronic medium with text format. Later on, data recorded between 2007–2010 is also added to the analysis. It also contains 'metadata' consists of stations' information regarding their activation year, location, active periods and a list of climatological variables recorded. As the first step, the dataset stored in two columns:

recording time and observation as a text format is converted into an Excel file as in the form necessitated by the DM functions.

There are two common problems of meteorological data that the climate researchers have to deal with before conducting any analysis [40]: (1) Missing observations, (2) Data inhomogeneity; otherwise, they have to restrict themselves to the complete and homogeneous data obtained from a very limited number of stations only (usually between 50–125 stations, which is at most 50% of the Turkish meteorological data). Nevertheless, we should keep in mind that climate change related studies require long series of data.

After checking for homogeneity and imputing the missing values, standardization (i.e., normalization) is applied to the climate variables before using any DM functions. Note that this approach is very common in multivariate statistical analysis mainly because some variables may have disproportionate effects on the results due to their different measurement scales.

6.3.1 Check for homogeneity of data

The quality of meteorological data should be controlled to obtain reliable analysis results. An important such control is the test for homogeneity of the dataset. Inhomogeneity is described as "changes in the series due to non-climatic reasons;" these factors may prevent the detection of true climatic patterns which may lead to bias results [16]. For that reason, temporal homogeneous data series are needed in climate related research.

Non-climatic reasons, which cause nonhomogeneity, include relocation of a station, changes or any breakdown in the instruments, changes or modifications in the calculations, or urbanization around the stations [30]. For example, Fethiye and Antalya stations were relocated in 1962, and therefore, the series obtained from these stations are inhomogeneous [35]. These reasons can lead to abrupt, gradual (or instant) changes or breakpoints (change points), which are observed as change in the mean, the variance, or in both, or change in trend of the data. Thus, it is essential to detect and correct inhomogeneity that may exist in the series before conducting any analysis on the meteorological stations' data. Availability of historical stations' data (i.e., metadata) helps the researchers to solve this problem easily. Unfortunately, recording such useful data is usually ignored at the stations most of the time. As a result, statistical homogeneity tests are conducted on the series to decide objectively if there exists inhomogeneity or not.

Homogeneity tests used in the literature can be classified into two groups: absolute tests and relative tests. While the absolute tests are conducted within the series, relative tests depend upon the relationship between neighbor stations [43]. Many of such tests used in the previous climate related studies assume the independence of the observations. However, meteorological data is a time series and autocorrelation between data is unavoidable. As a result, the power of the homogeneity tests, which assume the independence of observations is questionable [42].

Realizing the importance of checking for data homogeneity, the NINLIL Climate Research Group starts with questioning the power of the homogeneity tests used in literature, and also, the ones not used at all but possibly can be considered for this purpose. The very first study conducted in this framework by NINLIL is related to the validity of the absolute homogeneity tests for meteorological time series data [42]. To achieve this aim, first, a time series model representing the Turkish monthly average temperature variable recorded in the period from 1950 to 2006 is built. Then, various inhomogeneity scenarios for low and high variability are created.

(1) to test for a mean shift at the beginning, in the middle, and at the end of the series for representing any change in the location of the station for example;

(2) to test for gradual six years change at the beginning, in the middle, and at the end of the series for representing the instrument change case;

(3) to test for a sharp decrease in temperature at a one time point for representing a sudden change case;

(4) to test for the trend for representing the change in the surroundings of the station. In this study, Kruskal-Wallis (KW), Friedman, transpose Friedman, and KPSS tests are all employed. Here, the Friedman test is particularly selected since it considers data dependency. Results show that although the KW test reveals better results than Friedman, none of the methods considered outperforms the others. Following, a more realistic simulation environment is conducted which can handle both the change in mean and the variance of the Turkish precipitation data recorded in the period 1950–2006 [15]. Methods include Friedman test, Cumulative Sum (CUSUM) and Shewart Control Charts (SCC). Results are also compared with that of the Standard Normal Homogeneity Test (SNHT) obtained from another study. These tests have certain dis/advantages to apply to the climate data. To illustrate, only the Friedman test considers dependency among the others. While CUSUM and SCC can test inhomogeneity both due to mean and/or variance shift, Friedman and SNHT can only detect changes due to the mean shift. Besides, SNHT, CUSUM and SCC require normality of observations while Friedman does not. Only SNHT is a relative test that requires homogeneous neighbor stations but it may come up with non-testable results otherwise.

According to the simulation results, although the Friedman test considers data dependency, it does not perform very well, and also, cannot detect the exact location of the breakpoint if there is one. The SCC can detect small shifts due to both mean and variance. On the other hand, CUSUM provides information for the stations for which the SNHT test produces the non-testable or inconsistent results. Thus, it can be said that CUSUM and SCC are two promising homogeneity tests, yet they still cannot handle data dependency. Based on the study findings, two alternative methods, modified CUSUM and EWMAST, are suggested to be used to overcome the problems due to the independence assumption, and also, to consider variance shift.

Yet, in a following comprehensive study, the performance of the extensively used homogeneity test, SNHT, is compared to those of the ones which are not known that much in the domain by Monte Carlo simulation [43]. For this purpose, data is generated by using the normally distributed temperature variable. Both absolute tests such as KW, Friedman, Buishand range, Pettit, Von Neumann ratio, KPSS, ADF, and also, the relative tests such as GAHMDI, Bayesian change point analysis, F-test for structural breaks (i.e., Chow test), SNHT, RHT, CauMe (HOMER), TPR, are included in the comparisons. Although not much is used in the literature, note that ADF, KPSS, and Friedman tests are the valid ones for conducting homogeneity analysis. According to the simulation results SNHT, F-test, GAHMDI, and the RHT (i.e., the relative ones) have the best detection rates among the others. But if proper reference series do not exist, then the F-test can be used instead. Moreover, because KW, Friedman, ADF, Pettit, Von Neumann ratio tests may not produce reliable results, they should be used with caution. As a future study, the performance of the tests considered should also be evaluated under non-normality and in the existence of high autocorrelation.

As stated before, in literature, the SNHT test is the most extensively used one among the others. Nevertheless, in case the reference series are not highly correlated ($\rho < 0.8$) with the tested one, it fails to come up with a solution. In another study, an improvement on this test, called SNHT-Bootstrap (SNHT-BS), is made to overcome this difficulty by Yazıcı et al. [13]. In this test, a moving block bootstrap (MBB) which considers data dependency is utilized to build an empirical distribution of the test statistic. To evaluate the proposed method's performance, first, an exponential smoothing model is fitted to the 60 years of monthly temperature data obtained from 244 stations of Turkey. Next, its results are compared with that of the F-test. Under the two scenarios: the mean shift and sudden decrease, the proposed method works well when compared to the F-test, especially, for the breaks located in the middle of the series. However, it still needs to be improved to better capture the Type- I Error probabilities.

Recently, a new absolute homogeneity test is developed to overcome the above mentioned disadvantages by using a computational statistics approach [41]. In this test, called LRT-BS, the likelihood ratio test for mean shift in AR (p) models is incorporated with MBB to detect single/multiple breakpoints, especially, at the beginning and at the end of the series where inhomogeneity detection is problematic. The test statistic is derived by using exact likelihood, and its critical values are obtained for different samples and block sizes using simulation. Performance of this new method is compared to those of SNHT and F-test. Results show that LRT-BS performs better than the other two. All three tests are also applied to the temperature and precipitation series of the Fethiye station of Turkey, which was moved to elsewhere in its vicinity in 1962. Only the LRT-BS test correctly detected the existent breakpoint in the series. This approach is going to be improved to be applied to the variance changes as well. As a future study, all stations in Turkey are going to be checked for homogeneity using these newly developed and more powerful tests.

6.3.2 Missing data handling

It is very common to have missing observations in meteorological datasets collected from stations due to several reasons like broken instruments, extreme meteorological events, and recording errors. Recently, TSMS has about 270 actively working stations located all around Turkey. Because the quality checks were not done, data collected before 1975 is not published. Besides, most of the series have about 50% missingness from 1950 to 1960, and they have not been included in the analysis. Furthermore, some series are not included at all due to their inhomogeneity. On the other hand, sound climatic analysis requires the study of lengthy time series representing the region to come up with realistic and useful outcomes. Hence, this leads to reviving the missing value imputation research.

The very first step in handling missing observations in meteorological series is to identify missing observations. To illustrate, in our dataset, the monthly total precipitation series contains no values (i.e., left as blank) that caused such an analysis to be conducted. An investigation on the matter reveals that TSMS records the amount of precipitation as zero if it is less than or equal to 0.1 mm but lefts as blank if there is no precipitation at all. On the other hand, if an observation is missing (i.e., not recorded), there is also no value entered (i.e., left blank) into the dataset. Consequently, there is confusion about missing values and no precipitation at all. Therefore, a meticulous discriminative analysis is conducted to identify if an observation is really missing or the amount of precipitation is below the significant lower bound, or there is no precipitation at all. In this case, the station's metadata, the neighboring stations, temperature series, and the values of the same observation month in different years are checked. If in the end it is determined that there is no precipitation at all for that month, the precipitation amount is recorded as zero. Otherwise, it is labeled as missing observation. As a result of this analysis, 18% of the precipitation data, which is a considerable amount, is recovered.

In literature, the most preferred approach of filling these observations out is to use the mean value imputation due to its simplicity. However, more elaborate approaches are needed for truly representing the spatio-temporal properties of the series. To this end, a study is conducted by the NINLIL Research Group to compare the performances of missing value imputation methods in time series [6, 40]. Six commonly used methods included in this study are simple methods such as Simple Arithmetic Average (SAA), Normal Ratio (NR), and NR Weighted with Correlations (NRWC), and computationally intensive methods such as Multi-Layer Perceptron Neural Networks (MLPNN) and Expectation-Maximization Monte Carlo Markov Chain (MCMC-1), and an improved version of it (MCMC-2). Note here that while imputing meteorological time series, the missing at random (MAR) approach is a plausible assumption.

Monthly total precipitation and monthly mean temperature variables recorded from 1965 to 2006 (40 years) are used in this study. To be able to represent spatial features of the series as well, seven basins from seven climate regions of Turkey, namely South Eastern Anatolia (SAN), Eastern Anatolia (EAN), Mediterranean (MED), Central Continental Anatolia (CCN), Aegean (AEG), Marmara (MAR), and Black Sea (BLS), are selected. Note that these have long been accepted as the climate regions of Turkey. From each region, a station with no missing values for the period considered, and also, surrounded by highly correlated reference stations is selected as the target station. When examined carefully, it is observed that the Turkish Meteorological database contains long-term missing periods corresponding to 3–6 years or more. So that, in the simulation, three different types of missingness scenarios consisting of four (i.e., 10% of missingness), eight (i.e., 20% of missingness) and 21 (i.e., 50% of missingness) years of missing periods are studied. In the study, three types of artificially created missingness in both variables of seven selected targets are imputed by using the methods listed above. In their performance evaluations, besides graphical analysis, coefficient of variation for RMSE (CVRMSE) and correlation dimension (CD) measures are utilized. Here, CD is a novel measure for the performance evaluations of the missing data imputations in the meteorological series, which evaluates the series for its dispersion (Table 6.1).

According to the results, although it is easy to apply the simple imputation methods, they have commonly low efficiencies. The NR method performs better when compared to the other two. On the other hand, MCMC methods outperform and produce the most robust results overall for all missingness scenarios and regions for both measures (Table 6.1). In the existence of highly correlated reference stations ($\rho > 0.80$), imputations produce more reliable results as it is observed in the İzmir station selected as the target of the AEG region.

Table 6.1: Performances of the imputation methods studied (mean ± standard deviation).

Variable	**Precipitation**		**Temperature**	
Methods	**CVRMSE**	**CD**	**CVRMSE**	**CD**
SAA	0.533±0.187	0.212±0.175	0.121±0.185	0.096±0.131
NR	0.498±0.140	0.097±0.110	0.077±0.099	0.088±0.141
NRWC	0.529±0.186	0.140±0.153	0.121±0.184	0.092±0.133
MLPNN	0.445±0.105	0.081±0.109	0.057±0.059	0.060±0.103
MCMC-1	0.437±0.099	0.109±0.097	0.051±0.049	0.046±0.083
MCMC-2	0.437±0.099	0.072±0.096	0.051±0.049	0.069±0.108

6.4 Climate Data Mining

6.4.1 Determine if the climate has changed by descriptive mining summarization

Descriptive mining is applied to each meteorological variable to pinpoint the characteristics of the climate data. It also helps us generate various hypotheses that will lead us to conduct new research. As an example, examination of the outlying observations (e.g., extreme precipitation) and of their changes over time, can help us to investigate global climate changes effects on Turkey [4]. On the other side, examination of precipitation characteristics of the commonly accepted climate regions will give us hints about the change in the precipitation and the climate regions [18, 19, 20]. Moreover, outlying observations determined by the descriptive statistics will help us to develop early warning systems to determine natural disasters which may happen upcoming years beforehand [11].

In this section, descriptive analysis of the precipitation data collected between 1950–2006 will be exemplified [3]. Because it helps us to examine the data more realistically, first, monthly precipitation series are formed for all 277 stations. Then, descriptive analyses of the series are obtained concerning the rural/urban area, the stations, the commonly accepted climate regions and months. When analyzed according to the regions, while the maximum average precipitation amount (70.3 kg/m^2) is observed in the BLS region, the minimum average precipitation amount (34.8 kg/m^2) is observed in the SEA region. While average precipitation amount has the maximum spread with 71.61 kg/m^2 and 70.20 IQR (interquartile range) in the EAN; its spread in the BLS, MAR, EGE, EAN, and SEA regions are similar with respect to the IQR values. Series are all right-skewed, which means higher precipitation amounts than their means are observed more frequently than that of their lower precipitation amounts. Furthermore, having the minimum spread for both standard deviation (53.92 kg/m^2) and IQR (54.70 kg/m^2), it can be said that average precipitation amount is more homogeneous in the AEG regions than others.

On the other hand, the average monthly precipitation totals falling in rural and urban areas are 52.5 kg/m^2 and 51.9 kg/m^2, respectively. Moreover, when total precipitation amounts are examined, the maximum amount (309.5 kg/m^2) falls Trabzon-Hopa station in November while the minimum amount (0.032 kg/m^2) falls Diyarbakır-Nüsaybin station in August. The most spread series with the standard deviation of 168.2 kg/m^2 is observed in the Antalya-Center station.

When the average precipitation data is analyzed monthly (Table 6.2), the maximum amount falls in Turkey in December (90.95 kg/m^2) while the minimum amount falls in August (15.04 kg/m^2). The distribution of the series observed monthly is also a right-skewed distribution. Maximum and minimum deviations are observed in December and in June to August, respectively. The lower quantile values are zero for July and August because of no rain at all. The maximum average precipitation amount is observed in November, therefore, this

Table 6.2: Descriptive statistics of average precipitation amounts (kg/m^2) with respect to months.

Month	Mean Prec.	Std. Dev.	$Q_{0.25}$	Median	$Q_{0.75}$	Max	IQR
January	81.36	76.09	30.70	60.10	110.10	797.80	79.40
February	70.09	56.96	30.80	54.00	93.70	624.60	62.90
March	65.68	47.45	31.90	54.10	87.20	398.80	55.30
April	60.29	39.94	31.80	52.00	78.70	395.60	46.88
May	48.86	38.44	22.00	41.50	67.25	669.20	45.25
June	29.93	32.87	6.00	21.30	43.90	363.30	37.90
July	15.60	28.87	0.00	4.90	19.70	397.90	19.70
August	15.04	36.48	0.00	2.60	15.90	638.90	15.90
September	23.80	41.82	1.20	9.50	29.10	456.00	27.90
October	53.87	57.04	17.70	38.30	70.85	560.80	53.13
November	72.72	64.90	28.60	55.60	97.35	907.20	68.75
December	90.95	81.06	37.70	64.90	123.40	704.70	85.70

month can be considered as the one where more floods will possibly occur in the future.

Clustering

To answer the critical question "Has the climate been changing in Turkey?" posed in Section 6.1, a novel approach is adopted where a comparative statistical analysis is conducted in combination with clustering in two consecutive periods, 1950–1980 and 1981–2010. In our leading study, only the precipitation data collected from 247 stations in the period 1950–2010 are considered [18]. All series are assumed to be homogeneous and the missing values are imputed with the MCMC method. Using the two well-known clustering methods: k-means and hierarchical clustering, six, seven, and eight clusters have been formed. The ultimate number of clusters is determined as six by using both objective evaluations of the elbow method, and also, the subjective evaluation of the experts in the domain. Results reveal that apparent indicators are pointing out regional climate change signals in Turkey.

Following, a similar but more rigorous and objective comparative statistical analysis is conducted again on two 30 years long consecutive periods using the same approach [36]. The main aim of this novel approach is to determine whether there are statistically significant differences in the long-term mean and variance of the meteorological variables studied by comparing these parameters in two consecutive periods of climate regions that emerged. Note here that at least three decades of data are required in each of the periods for determining if climate change is occurring, effectively. This time in addition to the monthly total precipitation, minimum, maximum, and mean air temperature series recorded at 244 meteorological stations operated by the TSMS are considered. Here, 33 series are left (out of 277) since the missingness in these series is greater than 50%.

Table 6.3: Descriptive statistics of the variables for the time periods 1950–1980 (before) and 1981–2010 (after) for all of Turkey.

	Maximum Temperature		Minimum Temperature		Mean Temperature		Total Precipitation	
Statistics	**Before**	**After**	**Before**	**After**	**Before**	**After**	**Before**	**After**
L-T Ave.	25.38	25.71	1.29	1.63	12.92	13.11	53.09	52.26
Std. Dev.	8.86	9.01	9.61	9.85	8.43	8.92	56.67	57.75
Min.	-16.40	-3.00	-45.60	-42.80	-21.30	-17.60	0.00	0.00
Max.	50.00	49.80	25.20	27.00	35.70	37.30	797.80	907.20
Kurt.	2.47	2.40	3.25	3.189	2.55	2.41	12.65	14.68
Skew.	-0.35	-0.32	-0.52	-0.44	-0.24	-0.20	2.41	2.56

In this comparative study, first, descriptive mining is applied using statistical (Table 6.3) and graphical methods. Then, k-means and hierarchical clustering (i.e., Ward) methods are applied to each variable, separately. While determining the proper number of clusters objectively, elbow and dendrogram approaches are used in k-means and Ward methods, respectively. Besides, we searched for commonality in both methods; in other words, the number of clusters is determined as the one which leads to similar patterns in both methods. As a result, we come up with the decision that seven clusters are more appropriate for the four variables studied. Due to the similarities in the results of both methods, we continue analysis with the hierarchical method's output mainly because it also helps us to form sub-regions of seven clusters where necessary. The pattern of the seven main regions observed as a result is almost the same in two consecutive periods with the exception that some regions are getting smaller while others are becoming larger.

To compare if there are statistically significant differences in the means and variances of the four variables studied, paired samples Student's t and Pitman-Morgan (P-M) tests are conducted in the same regions. Findings of the analysis can be listed as follows:

- Small increases and small decreases are observed in the averages of the temperature series and the averages of the total precipitation series, respectively, after 1980.
- Increases in the standard deviations of all variables are observed after 1980.
- While small changes in means are detected, the changes in variances are considerable, one of the important indicators of climate change.
- Especially, there exists a large increase in the maximum total precipitation in the second period (i.e., after 1980), which is a clear indication of the upcoming risks of extreme precipitation.

In the end, we conclude that the semi-arid and dry sub-humid CCA region's climate characteristics are expanding whereas the semi-humid and humid temperature coastal climate regions' characteristics are contracting in Turkey. These important changes will possibly lead to disasters due to drought, and flood risks in most of Turkey today and in the future. This result can be evaluated as a change in the climate in Turkey has already been started.

As a future study, more climate variables including atmospheric humidity, cloudiness and so on can be included in a similar analysis with different clustering techniques such as fuzzy clustering and kernel methods.

6.4.2 Clustering climate regions

During the preliminary studies, the best method for clustering the climate regions of Turkey is searched. In one of the early studies, before the clustering analysis is conducted, missing value imputation is applied as follows [20]. Series having more than successive two years long missing values are not considered at all. Mean imputation is applied for the missing values recorded for months within a year. As a result, only 65 stations' data consisting of seven temperature variables (i.e., minimum and maximum temperature, average temperature, and average of maximum and minimum temperature, minimum and maximum of average temperature) collected in the period 1950–2006 are covered in the analysis. Here, two center-based clustering methods: k-means and fuzzy k-means are studied. Different clusters are obtained when each temperature is analyzed separately. But only five clusters are formed over Turkey when all variables are analyzed altogether. Both methods form the clusters in the same way except for two stations in the EAN. In addition, the AEG, MED and SEA regions have the same climatological characteristics. Except for the Trakya region, the MAR region shows BLS climate characteristics.

We suspect that the climate regions formed in the previous study may not represent the real situation because we only consider the temperature variables which do not represent spatial features of the stations' locations. Therefore, in a following study, in addition to the above variables, we also include the total precipitation variable in the analysis [19]. Besides, the variables are normalized before clustering the data using four different hierarchical methods: single-linkage, complete-linkage, average-linkage and Ward methods. The performance of methods are compared with respect to their stability and meaningfulness by applying the clustering methods on the variables obtained in two different time spans: 1950–2006 and 1970–2006. It is concluded that the Ward method provides the best results according to the criteria considered, and Turkey has seven climate regions where both temperature and precipitation variables are studied. However, the boundaries of these seven regions are different than the ones previously known. The climate characteristics of the CCA region is spreading in the country whereas the coastal climate region characteristics are contracting. This result emphasizes the fact that drought is expanding in the climate of Turkey. Findings of this study support the ones obtained in the previous study [20]. Nevertheless, more regions are found here mainly because of the precipitation variable. Common findings of both studies are:

(1) Stations located in the AEG region have similarities with the ones located in the MED region concerning the temperature variables.

(2) Stations in the MAR region have commonalities with the ones in the BLS region excluding those located in the Trakya part.

Finally, using all experiences gained from previous studies, an extensive clustering analysis leading to the most reliable results has been conducted [26]. This is an extension of the study by [19], where we apply the Ward method to nine climatological variables including seven air temperature series, total precipitation and relative humidity observed over the period 1970–2010. In addition to the data recovered in the precipitation series by discriminative analysis, all missing values of the nine variables (for at most 50% missing series) were imputed using the results of the studies conducted by NINLIL [5, 40], and hence, we could use the data from 244 stations out of 277. To the best of our knowledge, this is the first study where this big spatio-temporal data (i.e., stations and variables) have been used in applying the clustering analysis on the Turkish climate data.

Finally, it is decided that 12 (or 14 including two more sub-regions) clusters represent the climate of the region realistically (Figure 6.1). Table 6.4 displays the region names as well as the number of stations involved in each cluster. We believe that the resultant clusters reflect the physical and geographical characteristics of Turkey; so results of this study can be considered as a reference for the other researches of Turkey related to the climate because of the involvement of a long term course dataset having many meteorological variables.

In addition to applying the available methods to the data, we also developed a methodology which combines Dynamic Time Warping (DTW) with the con-

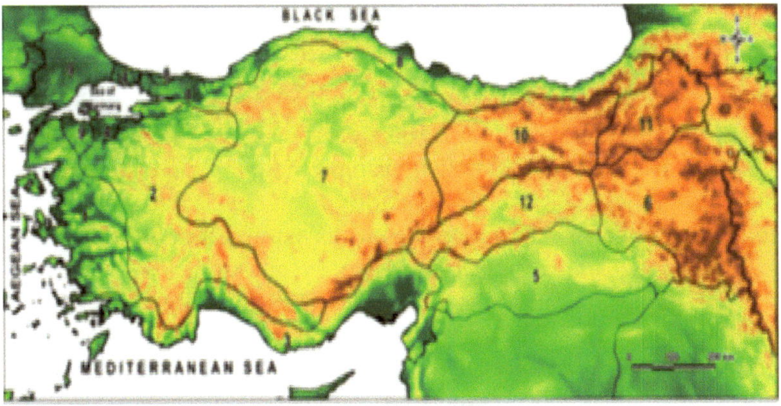

Figure 6.1: Climate regions of Turkey defined by Ward's clustering method.

Table 6.4: Region descriptions associated with the cluster numbers.

Cluster	Region Description	#Station
1	Dry Summer Subtropical Semihumid Coastal Aegean	25
2	Subhumid Mid-Western Anatolia	32
3	Dry Summer Subtropical Humid Coastal Mediterranean	25
4&9	Semihumid Marmara Transition	20
5	Dry Summer Subtropical Semihumid/Semiarid Continental Mediterranean	18
6	Semihumid Cold Continental Eastern Anatolia	18
7	Dry-Subhumid/Semiarid Continental Central Anatolia	52
8	Mid-Latitude Humid Temperate Coastal, Black Sea	18
10	Semihumid Continental Central to Eastern Anatolia	11
11	Rainy Summer Semihumid Cold, Continental Northeastern Anatolia	8
12	Semihumid Continental Mediterranean to Eastern Anatolia Transition	17

sensus clustering algorithm [39], which tries to combines the clusters formed by different methods into one. As a distance measure, DTW is relatively new and can handle out of phase series or series in different lengths or frequencies. DTW aligns the two time series in such a way that the distance between them is minimized. Nevertheless, it is used for a single variable and with standard clustering methods only. In this study, we try to take the advantage of using both consensus clustering and the DTW measure for developing an algorithm for multivariable applications. The performance of the algorithm is tested with both computer simulation and using Turkey's long-term meteorological time series data between the years 1950–2010. Results indicate that when used with consensus clustering, DTW performs better than the well-known Euclidean distance measure unless the cost constraint matters.

Last but not least, initiated from the needs of this research, new methods are also developed for temporal clustering of multivariate time series by a group of NINLIL researchers [7].

6.4.3 Identifying seasons by clustering

Consensus clustering is a powerful technique that improves the quality and robustness of the clustering solutions by taking advantage of various methods [21]. The idea here is to disturb the data by using resampling methods, called bootstrapping [29]. After using this approach for clustering Turkish meteorological data to determine its recent climate regions as explained in Section 6.4.2, this powerful technique is also applied to answer the question posed in Section 6.1 "How many seasons do Turkey go through in a year recently? How can we define these seasons?" [39]. For this purpose, the consensus clustering method is applied to the same data collected from 244 stations on five different variables including monthly maximum temperature, minimum temperature, average temperature, average humidity and total precipitation over the time span 1950–2010. One can refer to Fahmi et al. [19] for the detailed data description.

Four different clustering algorithms namely agglomerative nesting (Agnes), divisive analysis (Diana), PAM and k-means are integrated into one. During the analysis, different number of clusters are tried. Two measures of consensus are used in the algorithm which tries to quantify the concentration of the consensus distribution (CDF). These are the area under the CDF (AU CDF) and Δ(k), which is the proportion increase in the CDF area. The goal here is to find the k, the number of clusters, which maximizes this concentration. So that we pick the largest k which causes a large enough increase in the area under the associated CDF.

Based on both the objective and the experts' evaluations on the results, it is decided that three or four clusters seems to better represent seasonal variations of Turkey (Table 6.5). When four clusters (k = 4) are considered, winter season covers December, January and February months as usual. An extended summer is defined by the months June, July, August and September. While April and May correspond to spring, and October represents fall. March and November are interpreted as the transition seasons lying between standard regions. When, on the other hand, three clusters (k = 3) are considered, it can be said that Turkey is going through longer winter and summer seasons, and rather short spring and fall of a months duration.

Table 6.5: Results of consensus clustering for k = 3 and 4.

	Agnes		**Diana**		**PAM**		**k-means**		**Merged**	
k =	3	4	3	4	3	4	3	4	3	4
January	1	1	1	1	1	1	1	1	1	1
February	1	1	1	1	1	1	1	1	1	1
March	1	2	1	2	1	2	1	2	1	2
April	2	3	1	3	2	3	2	3	2	3
May	2	3	2	3	2	3	2	3	2	3
June	3	4	3	4	3	4	3	4	3	4
July	3	4	3	4	3	4	3	4	3	4
August	3	4	3	4	3	4	3	4	3	4
September	3	4	3	4	3	4	3	4	3	4
October	2	3	2	3	2	3	2	3	2	3
November	1	2	1	2	1	2	1	2	1	2
December	1	1	1	1	1	1	1	1	1	1
AUCDF	0.75	0.83	0.73	0.82	0.73	0.82	0.64	0.62	0.71	0.77
$\Delta(k)$	0.38	0.10	0.30	0.13	0.30	0.12	0.16	-0.04	0.28	0.09

6.4.4 Precipitation modeling

Once the climate regions of Turkey are defined, the next step is to develop early warning systems for upcoming drought and floods. NINLIL Climate Research Group is conducted several studies for modelling precipitation. These can be categorized into three groups with respect to the response variable of interest:

(1) Precipitation amounts (e.g., [8, 10, 27, 31, 38, 44]) (2) Absence/presence of precipitation (i.e., occurrence) (e.g., [14, 37]) (3) Extreme precipitation (e.g., [2, 4]). Among them, various models are developed using Linear and/or Nonlinear Time Series Models (TSM) (e.g., [8]), Time Series Regression Models (TSRM) (e.g., [9, 38]), Hidden Markov Models (HMMs) (e.g., [12]), MARS (e.g., [27]), CMARS and RCMARS [31], for certain critical regions such as CCA or for all over Turkey (e.g., [27]).

An example for the first group of studies, consider the one in which models are tried to be developed for precipitation prediction, particularly for the CCA region, where severe drought has been experienced for the last few decades [31]. To achieve this aim, seven meteorological variables namely monthly total precipitation, monthly temperature, monthly relative humidity, cloudiness, vapor pressure, surface air temperature and mean pressure collected from 43 stations of the CCA region of TSMS over the period 1976–2010 are used. Note here that stations are selected in accordance with our previous climate region determination study [26]. In addition to seven variables, two more variables mixing ratio (0.622 vapor pressure/pressure-vapor pressure) and time are also included in the models. The dataset is preprocessed by filling out the missing values using the MCMC method, and then, are also normalized. Besides, yearly aggregates (average) of each variable over all stations are calculated for the purpose of data reduction to obtain better fit of the model to the data. Next, MARS, Convex MARS (CMARS) and Robust CMARS (RCMARS) models are constructed and their performances are compared using the hold-out method in which the training and the test data are collected through 1976–2005 and 2006–2010, respectively, with respect to the precision (i.e., prediction variance), accuracy (i.e., R^2, MAE, RMSE, and r) and stability (i.e., comparing performances on both training and testing data) criterion.

According to the results, RCMARS led to the construction of the best precipitation prediction model being approximately twice as precise and accurate as the MARS and CMARS models are; it also has high stability when compared to those of the other two. Moreover, the RCMARS model builds can successfully be used for prediction purposes because it has a good fit to data with R^2 and r values being 79% and 90%, respectively. When this model is examined in detail, it is seen that the mean temperature, cloudiness, pressure and vapor pressure and their certain lags, to represent the time dependency, are the most significant variables which explain the variability in precipitation in this region.

Yet in another first group study, MARS, capable of handling nonlinearity in high-dimensional data is applied to the meteorological data collected from 244 stations located over Turkey during the period of 1950–2010 [27]. By considering the 12 clusters formed [26], total monthly precipitation is modeled for each cluster separately using seven air temperature series, relative humidity and cloudiness. Distribution of the precipitation amounts is given in Figure 6.2. Note here that, while developing models, despite the geographical proximity, clusters 4 and 9, show similar results, and thus, they have combined into one cluster.

For each cluster, training and testing datasets are selected to be 50 years from 1950 to 2009, and the last year, 2010, respectively. The PCA with lags approach where PCA for temperature series, other variables with the lags of them and lags of precipitation series incorporated in the model, provide a good fit to data (Figure 6.3).

To illustrate the second group of studies consider the one in which daily occurrence (and also the amount of precipitation) in the wettest (e.g., Eastern BLS), in one of the driest (e.g., CCA) and the normal moisture (e.g., AEG) regions of Turkey by using HMMs [33]. For this purpose, daily total precipitation series from 1964 to 2005 for the Eastern BLS region, from 1977 to 2006 for the CCA region, and from 1972 to 2005 for the AEG region are used. Three stations are chosen from each of the regions according to the closeness and correlation between them to conduct a successful local analysis.

It is seen that HMMs produce good results for the region with normal moisture climate to estimate daily precipitation occurrences, whereas, they perform better for the wettest region to estimate the daily precipitation amounts. They also provide useful insights for predicting the most probable states that represent the daily precipitation occurrence when a sequence of observations and the model parameters are known.

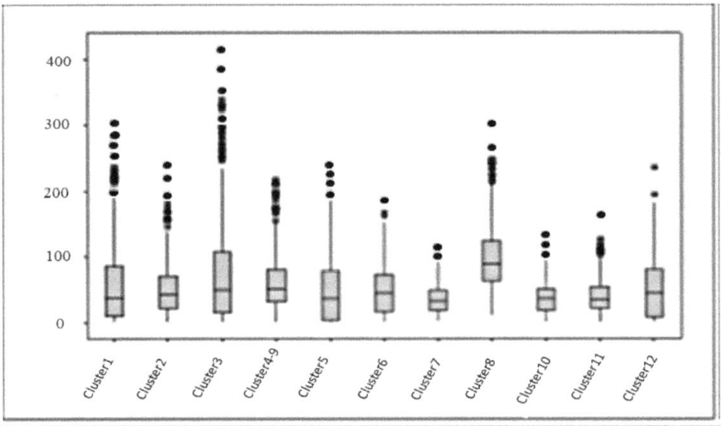

Figure 6.2: Boxplots of total monthly precipitation amounts.

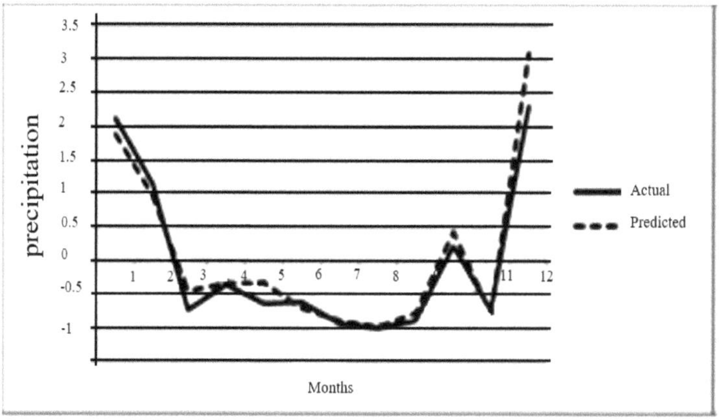

Figure 6.3: Plot of predicted versus actual values of test set for cluster 3 by using PCA and lags approach.

For exemplifying the third group of studies, let's focus on the extreme monthly precipitation events of Turkey for the period of 1950–2006 using 277 meteorological stations [4]. It is known that the precipitation variable is a right-skewed, especially, gamma type distributed variable. Hence, after applying a proper Box-Cox transformation on the monthly precipitation series for each station, we obtain an approximately symmetric distribution. Then, using whiskers of three times IQR, the extreme precipitation events for each month are determined. To see the possible existence of an increasing trend in the extreme precipitation events, the Augmented Dickey-Fuller (ADF) unit root test is applied, and it is concluded that the series is not stationary, and there is an increasing trend in the extreme precipitation events. A Box-Jenkins model, specifically ARIMA (2,1,0), is fitted to the stationary log-transformed series. After checking all the diagnostics, forecast values of the extreme events have been obtained. Results show that there is an increasing trend in the number of extreme events.

6.5 Conclusions and Future Studies

In this chapter, KDD and particularly DM approaches to the meteorological data are exemplified by the studies of NINLIL Climate Research Group, which consists of data scientists from various fields and has been active since 2006. The main aim of this research group is to investigate if climate change is taking place in Turkey, and if so, to determine its effects on the climate regions, and also on the seasons. The group is applied available methods to find answers to the research questions they posed in Section 6.1. Alternatively, they are improved or completely new ones are developed where necessary. The lessons learned from these studies can be briefly listed as follows:

- The KDD process is found to be very functional for analyzing climate data. Because objective evaluations of the results may not always come up with sound outcomes, the need for domain expert judgement is unavoidable for realistic interpretations. So from the Data Science's perspective, they are indispensable research partners for the success of DM applications.
- Meteorological data needs to be preprocessed carefully before DM applications because there exist a considerable amount of missing observations due to inactive stations or broken instruments, and also, inhomogeneous series due to the relocation of the stations. Because the 'metadata' simply detects inhomogeneity in series, we suggest that it should be kept in every station meticulously.
- SNHT and F-test are the two best tests for detecting inhomogeneity objectively where 'metadata' does not exist. To overcome their unrealistic assumptions, the SNHT-BS and LRT-BS methods are developed by the group. While SNHT-BS makes non-testable cases of SNHT possible to test, LRT-BS performs better than the SNHT and F-test for detecting the mean shift in series.
- Missing value imputation is a critical action to have a sufficiently long time series for reliable meteorological data analysis. Instead of using simple mean imputation, the MCMC method is suggested to be used for imputing the series having less than 50% missingness.
- To determine if the climate has changed or not, a novel approach in which two consecutive 30 years' time periods of the series are first clustered and then compared in the common clusters statistically is found to be very convenient.
- Individual clustering methods are very practical for detecting climate regions. However, to obtain better results, consensus clustering, which combines several individual methods' outputs into one, can be utilized. Moreover, a novel approach developed for temporal clustering of multivariate time series can be used instead.
- MARS, RCMAR, HMMs are very applicable techniques for developing prediction models for both occurrences and the amounts of precipitation.
- Clustering is also handy for identifying the seasons. Nevertheless, by treating yearly data as circular data, an association algorithm like apriori can be utilized for the search for better results.

Similar analyses and developments will be conducted with more meteorological variables and the use of up-to-date datasets from TSMS to continue following ongoing changes in the climate of Turkey.

Acknowledgement

The author would like to express her gratitude to all members of the NINLIL research group (http://stat.metu.edu.tr/en/ninlil-research-group) for their contributions. The climate research carried out by the NINLIL research group is supported by Middle East Technical University (METU), Ankara, Turkey, under the contract numbers BAP-2008-01-09-02, BAP-01-09-2015-001, and BAP-01-09-2014-001.

References

[1] C.C. Aggarwal. *Data Mining: The Textbook*. Springer New York, 2015.

[2] B. Aksoy, V. Purutçuoğlu, İ. Batmaz and C. Yozgatlıgil. Modeling the extreme precipitation data: Case study from Turkey. *26th European Conference on Operational Research-INFORMS.* Rome, Italy, July 1–4, 2013, 221.

[3] Ö. Asar, E. Kartal-Koç, S. Aslan, M.Z. Öztürk, C. Yozgatlıgil, İ. Çınar, İ. Batmaz, G. Köksal, M. Türkeş and H. Tatlı. Analysis of Turkey precipitation data from 1950 to 2006 using descriptive data mining techniques (in Turkish). *İGS'2010: 7. Statistics Days Symposium*, pp. 77–79. Ankara, Turkey, June 28–30, 2010.

[4] Ö. Asar, C. Yozgatlıgil, E. Kartal and İ. Batmaz. Analysis of extreme precipitation events in Turkey. *7. International Statistics Congress (ISKON)*. Antalya, Turkey, pp. 164–165, April 28–May 1, 2011.

[5] S. Aslan, C. Yozgatlıgil, C. İyigün, İ. Batmaz and H. Tatlı. Performance comparison of missing data imputation methods in meteorological time series using correlation dimension analysis (in Turkish). *Journal of Statistical Research*, 8(2): 55–67, 2011.

[6] S. Aslan. Comparison of missing value imputation methods for meteorological time series data. MSc Thesis, Department of Statistics, Middle East Technical University, Ankara, Turkey, 2010.

[7] S. Aslan, C. Yozgatlıgil and C. İyigün. Temporal clustering of time series via threshold autoregressive models: Application to commodity prices. *Annals of Operations Research*, 260(1-2): 51–77, 2018.

[8] S. Aslan, C. Yozgatlıgil, C. İyigün and İ. Batmaz. Analyzing the long-term projections of precipitation totals via linear and nonlinear time series models. *ICACM-International Conference on Applied and Computational Mathematics*. Middle East Technical University (METU), Ankara, Turkey. October 3–6, 2012.

[9] S. Aslan, C. Yozgatlıgil, C. İyigün and İ. Batmaz. Determining different regimes using nonlinear time series with a threshold value model: An application on Turkey precipitation data (in Turkish). *YAEM: 35th National Congress of Operations Research and Industrial Engineering*. Ankara, Turkey, September 9–11, 2015.

[10] F. Aykan, E. Kartal-Koç, C. Yozgatlıgil, C. İyigün, V. Purutçuoğlu-Gazi and İ. Batmaz. Developing precipitation models for continental central Anatolia, Turkey. *25th European Conference on Operational Research*. Vilnius, Lithuania, 8–11 July, 2012, 208.

[11] İ. Batmaz and G. Köksal. Overview of knowledge discovery in databases process and data mining for surveillance technologies and EWS. pp. 1–30. *In*: A.S. Koyuncugil and N. Özgülbaş (eds.). *Surveillance Technologies and Early Warning Systems: Data Mining Applications for Risk Detection*. Hershey, PA: IGI Global Publisher (Idea Group Publisher), 2011.

[12] İ. Batmaz, N. Sivrikaya, C. Yazıcı and C. Yozgatlıgil. Precipitation prediction by hidden Markov models. *IFORS 2014: 20th Conference of the International Federation of Operational Research Societies*. Barcelona, Spain, 13–18 July, 2014.

[13] C. Yazıcı, İ. Batmaz and C. Yozgatlıgil. A bootstrap application on the standard normal homogeneity test (SNHT) when there is not highly correlated reference series. *Proceedings of 60th ISI World Statistics Congress*, Rio de Janeiro, Brazil, 26–31 July, 2015b.

[14] İ. Batmaz, N. Yaman and C. Yozgatlıgil. Modeling daily occurrence of precipitation for certain regions of Turkey using hidden Markov models (HMMs). *27th European Conference on Operational Research*. Glasgow, UK, 12–15 July, 2015.

[15] C. Yazıcı, V. Purutçuoğlu, C. Yozgatlıgil, K. Bayramoğlu, C. İyigün and İ. Batmaz. *Homogeneity Analysis of Turkish Climate Data*. METU-STAT-Technical Report-2012-001, March, 2012.

[16] A.C. Costa and A. Soares. Homogenization of climate data: Review and new perspectives using geostatistics. *Mathematical Geosciences*, 41(3): 291–305, 2009.

[17] M.H. Dunham. *Data mining: Introductory and Advanced Topics*. Pearson Education India, 2006.

[18] F. Fahmi, E. Kartal, C. İyigün, C. Yozgatlıgil, İ. Çınar, S. Aslan, M.Z. Öztürk, M. Türkeş and İ. Batmaz. Studying the effect of climate change on Turkey by clustering methods (in Turkish). *Proceedings of Geography Conference with International Participation*. İstanbul, Turkey, 7–10 September, 2011a.

[19] F. Fahmi, E. Kartal-Koç, C. İyigün, M. Türkeş, C. Yozgatlıgil, V. Purutçuoğlu, İ. Batmaz and G. Köksal. Identifying climate regions of Turkey using hierarchical clustering method (in Turkish). *Journal of Statistical Research*, 8(1): 13–25, 2011c.

[20] F. Fahmi, E. Kartal, C. İyigün, M. Türkeş, C. Yozgatlıgil, V. Purutçuoğlu, İ. Batmaz and G. Köksal. Determining the climate zones of Turkey by center-based clustering methods. pp. 171–178. *In*: J.A. Tenreiro Machado, D. Baleanu and A.C.J. Luo (eds.). *Nonlinear and Complex Dynamics: Applications in Physical, Biological and Financial Systems*. New York: Springer, 2011b.

[21] J. Ghosh and A. Acharya. Cluster ensembles. *Wiley Interdisciplinary Reviews: Data Mining and Knowledge Discovery*, 1(4): 305–315, 2011.

[22] P. Giudici. Applied Data Mining: *Statistical Methods for Business and Industry*. John Wiley & Sons, 2005.

[23] J. Han, M. Kamber and J. Pei. *Data Mining: Concepts and Techniques*. Morgan Kaufmann Publishers, 2012.

[24] D. Hand, H. Mannila and P. Smyth. *Principles of Data Mining*. MIT Press Massachusetts, 2001.

[25] I.H. Witten, E. Frank, M. Hall and C.J. Pal. *Mining: Practical Machine Learning Tools and Techniques*. Morgan Kaufmann, MA Cambridge, 2017.

[26] C. İyigün, M. Türkeş, İ. Batmaz, C. Yozgatlıgil, V. Purutçuoğlu-Gazi, E. Kartal-Koç and M.Z. Öztürk. Clustering current climate regions of Turkey by using a multivariate statistical method. *Theoretical and Applied Climatology*, 114(1): 95–106, 2013.

[27] R. Mansyrov, C. Yazıcı, C. Yozgatlıgil and İ. Batmaz. Statistical modeling in high dimensions with applications modeling of precipitation amount of Turkey by multivariate adaptive regression splines. *IC-SMHD-2016 —International Conference on Information Complexity and Statistical*

Modelling in High Dimensions with Applications, Nevşehir, Cappadocia, Turkey, May 18–21, 2016.

[28] B. Mirkin. *Core Concepts in Data Analysis: Summarization, Correlation and Visualization.* Springer Science and Business Media, 2011.

[29] S. Monti, P. Tamayo, J. Mesirov and T. Golub, T. Consensus clustering: A resampling-based method for class discovery and visualization of gene expression microarray data. *Machine Learning*, 52(1): 91–118, 2003.

[30] Meteorological Organization World (MOW) Guide to meteorological instruments and methods of observation: Part III—Chapter 3—Quality Management. WMO-No. 8, Geneva, 1996.

[31] A. Özmen, İ. Batmaz and G.-W. Weber. Precipitation modeling by polyhedral RCMARS and comparison with MARS and CMARS. *Environmental Modeling and Assessment*, 19: 425–435, 2014.

[32] D. Pyle. *Data Preparation for Data Mining*. Morgan Kaufmann, 1999.

[33] N. Sivrikaya. Modeling precipitation series of Turkey with hidden Markov models (HMMs). MSc Thesis, Department of Statistics, Middle East Technical University, Ankara, Turkey, 2015.

[34] T.F. Stocker, D. Qin, G.K. Plattner, M. Tignor, S.K. Allen, J. Boschung, A. Nauels, Y. Xia, V. Bex and P.M. Midgley. International Panel on Climate Change (IPCC), Summary for Policymakers. In: *Climate Change 2013: The Physical Science Basis. Contribution of Working Group I to the Fifth Assessment Report of the Intergovernmental Panel on Climate Change.* Cambridge University Press, Cambridge, United Kingdom and New York, NY, USA, 3–29, 2013.

[35] M. Türkeş, U.M. Sümer and İ. Demir. Re-evaluation of trends and changes in mean, maximum and minimum temperatures of Turkey for the period of 1929–1999. *International Journal of Climatology*, 22: 947–977, 2002.

[36] M. Türkeş, C. Yozgatlıgil, İ. Batmaz, C. İyigün, E. Kartal-Koç, F. Fahmi and S. Aslan. Has the climate been changing in Turkey? Regional climate change signals based on a comparative statistical analysis of two consecutive time periods, 1950–1980 and 1981–2010. *Climate Research*, 70: 77–93, 2016.

[37] N. Yaman, C. Yozgatlıgil and İ. Batmaz. Developing absence-presence models of precipitation data for certain regions of Turkey by hidden Markov models (in Turkish). *YAEM: 35th National Congress of Operations Research and Industrial Engineering.* Ankara, Turkey, September 9–11, 2015.

[38] C. Yazıcı, İ. Batmaz and C. Yozgatlıgil. The precipitation modeling of Aegean region of Turkey with multivariate adaptive regression splines and time series regression. *27th European Conference on Operational Research.* Glasgow, UK, 12–15 July, 2015a.

[39] A. Yetere-Kurşun. Ensemble clustering of Turkish meteorological data. MSc Thesis, Institute of Applied Mathematics, Middle East Technical University, Ankara, Turkey, 2014.

[40] S. Aslan, C. Yozgatlıgil, C. İyigün and İ. Batmaz. Comparison of missing value imputation methods for Turkish meteorological time series data. *Theoretical and Applied Climatology*, 112: 143–167, 2013.

[41] C. Yazıcı. A computational approach to detect Inhomogeneities in time series data. MSc Thesis, Department of Statistics, Middle East Technical University, Ankara, Turkey, 2017.

[42] C. Yozgatlıgil, V. Purutçuoğlu, C. Yazıcı and İ. Batmaz. Validity of homogeneity tests for meteorological time series data: A simulation study. *Proceedings of 58th ISI World Statistics Congress*, Dublin, Ireland, August, 2011.

[43] C. Yozgatlıgil and C. Yazıcı. Comparison of homogeneity tests for temperature using a simulation study. *International Journal of Climatology*, 36(1): 62–81, 2016.

[44] C. Yozgatlıgil, İ. Batmaz and N. Yaman. Modeling precipitation amounts for certain regions of Turkey by hidden Markov models (in Turkish). *YAEM: 35th National Congress of Operations Research and Industrial Engineering.* Ankara, Turkey, September 9–11, 2015.

III

OPERATION RESEARCH IN BUSINESS SCIENCE AND FINANCE

Chapter 7

Fundamentals of Market Making Via Stochastic Optimal Control

Emel Savku

7.1 Introduction

Market making has attracted the attention of scientists in a diversified field of research such as economics, finance, statistical physics, artificial intelligence, machine learning, and the main traditional path, mathematics within the framework of stochastic optimal control.

All the interests of the market participants; i.e., all the orders with their signs (buy or sell), the prices, the quantities, the exact time of the order recorded by the market (time-stamp), and different types of market based information are stored as a file in a central computer at an organized electronic market. Such a list of all possible transactions that can be performed on equity, a future, or a specified derivatives market at any given time is called a *Limit Order Book* (LOB). Buy and sell orders match on an LOB depending on a priority rule, which is based on the prices and at most of the markets, the arrival time of the orders is called First In, First Out rule. For a comprehensive review of the theoretical and the empirical mechanics, and nomenclature of an LOB, see [1, 16].

University of Oslo.

There are three main order types in an LOB:

- *Market Orders* (MOs): A buy or a sell order of a certain quantity, which is immediately executed at the best available price in the current market.
- *Limit Orders* (LOs): A buy or sell order to specify the limit price for a certain number of shares. A buy (sell) order can only be executed at the limit price or at a lower (higher) one. Hence, LOs may not be executed; actually, they are stored in the LOB until an MO fills or cancellation occurs. By the way, it is guaranteed that a trader does not pay more than a pre-determined price for an asset.
- *Cancellation order*: A cancellation of an existing limit order.

An LOB, see Table 7.1, has a fixed number of price levels and a minimum price movement. The gap between the price levels is called a tick size (for an assessment of a tick size plot program in the European financial market, see [23]). The best/lowest limit sell order is called the *ask* price and the best/highest limit buy order is called the *bid* price. The difference between bid and ask prices represents the bid/ask spread, which is always positive and a multiple of the market tick size. The mid-price of a stock is defined as the average of the bid and ask prices.

When we focus on the LOB or the market microstructure, we encounter mainly two types of agents in the financial markets: liquidity providers, who submit LOs, and liquidity takers, who submit MOs. Consequently, within the context of the LOB, a Market Maker (MM) may be a brokerage house, a firm or an individual, who works on the behalf of a large financial institution to provide trading services for investors. An MM continuously quotes bid and ask prices on the LOB and makes a profit by a bid/ask spread. Therefore, an MM is a liquidity provider for the financial market. Actually, with the help of electronic markets and high frequency trading, many agents can post LOs to the LOB at a specific price and a specified quantity at an *almost* fair playing field.

On the other side, any kind of trading activity carries various potential risks for the investors, and market making has no difference. Especially, an MM faces two main risks: the inventory risk and the adverse selection risk.

Table 7.1: A brief summary of an LOB with bid/ask prices, volumes, mid-price and price levels.

Quantity	Price	Asks
	103.00	24
	102.75	9
Mid-price	102.50	Mid-price
32	102.25	
7	102.00	
11	101.75	
Bids		Quantity

Inventory risk: An MM has to hold some amount of assets to provide dynamic liquidity opportunities to other market participants. It can not be ignored that the arisen inventory has a value concerning a benchmark or reference asset (e.g., dollar, euro), which is the current value of all assets held by MM. Any fluctuation in the market prices of the stocks will put pressure on the inventory value. For example, let us assume that a downward trend in asset prices starts in the financial market. Then, an MM may trade only on one side of the LOB. Her bid LOs will begin to be filled but not the ask orders, which will cause the trouble of accumulating inventory on the sell side. An MM's main purpose is to increase the value of her inventory. Hence, she would prefer a market atmosphere in which the prices are moving in a range and her bid/ask orders are being filled within an almost equal frequency. By the way, her inventory risk may be mitigated and her gain may be maximized via gathering incremental spread profits. At a downward trend period, an MM may stop all her operations and wait for a profitable price level. On the other hand, against obstinately dropping prices to prevent the risk of locking all assets on the sell side, she may prefer to sell at a lower price. Additional to the market trends, when we focus on the assets in the portfolio of an MM, the number of the stocks appreciating may be exploited and losing value ones may be unfilled. Therefore, an inventory risk may be faced because of buying and selling at the wrong times. Such phenomenons arise in stochastic optimal control literature as Profit and Loss (P&L) maximization problems for an MM. There are several works approaching inventory risk via volatility of the market and by representing the risk constraints in a utility function, for some of them, see [3, 8, 9, 17, 18, 19, 21, 32] and references therein.

Adverse selection risk: Actually in any market and any trading operation, one party may have greater material knowledge than the other party, which is called also asymmetric information or information failure. Hence, such a risk exists for MMs as well. Adverse selection risk arises, when a seller has more information than a buyer, and vice versa. These traders with private information act at the expense of liquidity providers, i.e., MMs. Generally, when MMs perceive an increase of information asymmetry, they do not prefer to cease providing quotes, instead, they widen the bid/ask spread to bear the increasing risk of buying or selling of an unfavorable asset. Moreover, they become more sensitive to the signals that they collect by observing investors' order flow. The microstructure literature provides several works investigating the adverse selection risk [8, 9, 6, 10, 29].

Furthermore, the pricing strategies and the risks encountered by such traders have been studied in microstructure extensively, for some surveys of examinations cf. [5, 27, 33].

This survey work is organized as follows: In Section 7.2, we introduce model dynamics leading the theory of market making via stochastic optimal control, especially, by the seminal paper [4]. In Section 7.3, we focus on inventory risk and present optimal quotes and related asymptotic properties by providing some results of [18]. In Section 7.4, we give an example of the adverse selection impact on an MM's trading strategy as explained in [8]. In Section 7.5, we study optimal bid/ask spreads of an MM in a European options market by expressing some results developed in [32]. The final section is devoted to a conclusion and an outlook of the development of the market making theory.

7.2 Model Dynamics

Avellaneda and Stoikov [4] proposed a novel model for market making in an LOB, which followed the theory developed by Ho and Stoll (1981) [20]. Ho and Stoll derived the bid/ask quotes around a *true* or *fair* price to take into account the impact of the inventory. Avellaneda and Stoikov (2008) [4] combined this framework with an intuitive two-step procedure close to the microstructure of an actual LOB based on the econophysics literature. For a given inventory, they derived the bid/ask quotes by computing the intensity of the execution of the quotes as a function of the distance of the prices from the mid-price.

The authors assume that the mid-price evolves according to:

$$dS(t) = \sigma dW(t), \qquad S(0) = s, \tag{7.1}$$

where $W(t),\ t \in [0,T]$ is a standard Brownian motion and $\sigma > 0$ is the volatility of the risky asset.

The goal of an agent is to maximize exponential utility of her P&L at a terminal time $T < \infty$. The agent sets the bid p^b and the ask p^a prices as follows:

$$\begin{aligned} p^b(t) &= S(t) - \delta^b(t), \\ p^a(t) &= S(t) + \delta^a(t), \end{aligned}$$

where δ^b and δ^a represent bid and ask spreads, respectively. The wealth and inventory of an MM depend on the arrival of MOs. At each time, when an LO of an MM is filled by an MO, there occurs a jump in her cash flow:

$$dX(t) = (S(t) + \delta^a(t))dN^a(t) - (S(t) - \delta^b(t))dN^b(t), \qquad t \in [0,T], \tag{7.2}$$

where $N^a(t)$ and $N^b(t)$ are the amount of the filled ask and bid orders, i.e., the number of the stocks sold and bought up to time t with the intensities λ^a and λ^b, respectively. Therefore, the number of the stocks held up to time t is

$$q(t) = N^a(t) - N^b(t), \qquad t \in [0,T].$$

In econophysics literature, the distribution of the market order size has been studied several times and it has been showed that it follows a power law as follows:

$$f^Q(x) \propto x^{-1-\alpha}$$

for a large x and with the changing α values of the stocks in the different financial markets, for details see [13, 15, 25].

However, various authors tried to describe the statistics of the market impact [13, 28, 35], actually, there is less consensus. Whereas, there is not an agreement how to define and measure it, Potters and Bouchaud [28] demonstrated that the change in the price Δp following a market order of size Q is given by

$$\Delta p \propto \ln(Q).$$

Hence, the Poisson intensity at which agent's limit orders are executed can be derived in terms of the distance δ between the mid-price and her quotes as follows:

$$\begin{aligned}
\lambda(\delta) &= \Lambda P(\Delta p > \delta) \\
&= \Lambda P(\ln(Q) > K\delta) \\
&= \Lambda P(Q > \exp(K\delta)) \\
&= \Lambda \int_{\exp(K\delta)}^{\infty} x^{-1-\alpha} dx \\
&= A \exp(-k\delta),
\end{aligned}$$

where Λ is a constant frequency of market buy or sell orders, $A = \Lambda/\alpha$, and $k = \alpha K$.

This formula can be explained intuitively: If the MM sets her quotes further away from the mid-price, then less often a market buy order may lift the agent's sell LO, and less often a market sell order may hit the agent's buy LO.

Finally, the goal of the MM is to maximize her P&L and to find a optimal bid/ask spread by solving the following stochastic optimal control problem:

$$u(t,s,x,q) = \max_{\delta^a,\delta^b} E\left[-\exp(-\gamma(X(T)+q(T)S(T)))\right]. \tag{7.3}$$

Several authors inspired by this seminal paper [4] developed related market making models. Additionally, Cartea et al. (2014) [9] considered the impact of MOs on the LOB and the effects of adverse selection risk. They proved closed form solutions of the optimal quotes by applying a first-order Taylor expansion. Moreover, Cartea and Jaimungal (2015) [7] investigated risk measures for high-frequency trading. Furthermore, Guilbaud and Pham (2013) [19] studied a model of market making not only with LOs but also allowing MOs. In the very comprehensive book of algorithmic and high frequency trading [8], many other theorems, illustrations, and extensions of this leading work, Avellaneda and Stoikov (2008) [4], can be found.

In the following section, we will provide the solution of the problem defined in Equation 7.3 under the inventory restrictions as described in [18].

7.3 Optimal Quotes under Inventory Risk and Market Impact

In this section, we focus on the market making problem under inventory limits by introducing the results of Guéant et al. (2013) [18].

In this work, authors assume that an MM with an inventory $Q > 0$ will not set a bid quote, and correspondingly, an MM with an inventory $-Q < 0$ will not set an ask quote. Here, risk limits are considered with a positive inventory and short selling is represented with a negative inventory. Furthermore, the intensities λ^b and λ^a associated with N^b and N^a respectively are defined as follows:

$$\lambda(\delta^b) = Ae^{-k\delta^b} = A\exp(-k(s-s^b)), \quad \lambda(\delta^a) = Ae^{-k\delta^a} = A\exp(-k(s^a-s)),$$

where A and k are positive constants that describe the liquidity of the stock. Any order is posted closer to the reference price (7.1), the faster it will be executed. Note that the cash flow of the MM is defined by Equation 7.2.

The solution process for the problem 7.3 is based on the classical techniques of the stochastic optimal control theory, especially, by the dynamic programming principle.

Authors define Hamilton-Jacobi-Bellman equations in the following way:
Let $q \in \{-Q, ..., Q\}$ for $(t,s,x) \in [0,T] \times \mathbb{R}^2$ be:
For $|q| < Q$:

$$\begin{aligned}
&\partial_t u(t,x,q,s) + \frac{1}{2}\sigma^2 \partial^2_{ss} u(t,x,q,s) \\
&+ \sup_{\delta^b} \lambda^b(\delta^b) \left[u(t,x-s+\delta^b,q+1,s) - u(t,x,q,s)\right] \\
&+ \sup_{\delta^a} \lambda^a(\delta^a) \left[u(t,x+s+\delta^a,q-1,s) - u(t,x,q,s)\right] = 0.
\end{aligned}$$

For $q = Q$:

$$\begin{aligned}
&\partial_t u(t,x,Q,s) + \frac{1}{2}\sigma^2 \partial^2_{ss} u(t,x,Q,s) \\
&+ \sup_{\delta^a} \lambda^a(\delta^a) \left[u(t,x+s+\delta^a,Q-1,s) - u(t,x,Q,s)\right] = 0.
\end{aligned}$$

For $q = -Q$:

$$\begin{aligned}
&\partial_t u(t,x,-Q,s) + \frac{1}{2}\sigma^2 \partial^2_{ss} u(t,x,-Q,s) \\
&+ \sup_{\delta^b} \lambda^b(\delta^b) \left[u(t,x-s+\delta^b,-Q+1,s) - u(t,x,-Q,s)\right] = 0.
\end{aligned}$$

with final condition:

$$\forall q \in \{-Q, ..., Q\}, \quad u(T,x,q,s) = -\exp(-\gamma(x+qs)).$$

In this setting, *CARA* utility functions allow us to represent Mark-to-Market value of the portfolio $(x+qs)$. Moreover, the exponential decay for the intensity functions, δ^a and δ^b, helps to reduce the corresponding Hamilton-Jacobi-Bellman equations in a form of ordinary differential equations (ODEs).

The following proposition brightens the path to reach the optimal control processes of this problem.

Proposition 7.1
(Guéant et al. (2013) [18]) Let us consider a family $(v_q)_{|q|\leq Q}$ of positive functions solution of:

$$\begin{aligned} \forall q \in \{-Q+1,...,Q-1\}, \quad \dot{v}_q(t) &= \alpha q^2 v_q(t) - \eta(v_{q-1}(t)+v_{q+1}(t)), \\ \dot{v}_Q(t) &= \alpha Q^2 v_Q(t) - \eta v_{Q-1}(t), \\ \dot{v}_{-Q}(t) &= \alpha Q^2 v_{-Q}(t) - \eta v_{-Q+1}(t), \end{aligned}$$

with $\forall q \in \{-Q,...,Q\}$, $v_q(T)=1$, where $\alpha = \frac{k}{2}\gamma\sigma^2$ and $\eta = A(1+\frac{\gamma}{k})^{-(1+\frac{k}{\gamma})}$. Then, $u(t,x,q,s) = -\exp(-\gamma(x+qs))v_q(t)^{-\frac{\gamma}{k}}$ is solution of (HJB).

The following proposition demonstrates the solution of these ordinary differential equations, which is a family of positive functions:

Proposition 7.2
(Guéant et al. (2013) [18]) Let us introduce the matrix M defined by:

$$\begin{bmatrix} \alpha Q^2 & -\eta & 0 & \cdots & \cdots & \cdots & 0 \\ -\eta & \alpha(Q-1)^2 & -\eta & 0 & \ddots & \ddots & \vdots \\ 0 & \ddots & \ddots & \ddots & \ddots & \ddots & \vdots \\ \vdots & \ddots & \ddots & \ddots & \ddots & \ddots & \vdots \\ \vdots & \ddots & \ddots & \ddots & \ddots & \ddots & 0 \\ \vdots & \ddots & \ddots & 0 & -\eta & \alpha(Q-1)^2 & -\eta \\ 0 & \cdots & \cdots & \cdots & 0 & -\eta & \alpha Q^2 \end{bmatrix}$$

where $\alpha = \frac{k}{2}\gamma\sigma^2$ and $\eta = A(1+\frac{\gamma}{k})^{-(1+\frac{k}{\gamma})}$.
Let us define

$$\begin{aligned} v(t) &= (v_{-Q}(t), v_{-Q+1}(t), \ldots, v_0(t), \ldots, v_{Q-1}(t), v_Q(t))' \\ &= \exp(-M(T-t)) \times (1,\ldots,1)'. \end{aligned}$$

Then, $(v_q)_{|q|\leq Q}$ is a family of positive functions solution of the Proposition 7.1.

Finally, by using a verification approach, authors present the value function and the optimal quotes of this *market making* problem.

Theorem 7.1

(Guéant et al. (2013) [18]) Let us consider $(v_q)_{|q|\leq Q}$ *as in Proposition 7.2.*

Then, $u(t,x,q,s) = -\exp(-\gamma(x+qs))v_q(t)^{-\frac{\gamma}{k}}$ *is the value function of the control problem.*

Moreover, the optimal quotes are given by:

$$s - s^{b*}(t,q,s) = \delta^{b*}(t,q) = \frac{1}{k}\ln\left(\frac{v_q(t)}{v_{q+1}(t)}\right) + \frac{1}{\gamma}\ln\left(1+\frac{\gamma}{k}\right), \quad q \neq Q,$$

$$s^{a*}(t,q,s) - s = \delta^{a*}(t,q) = \frac{1}{k}\ln\left(\frac{v_q(t)}{v_{q-1}(t)}\right) + \frac{1}{\gamma}\ln\left(1+\frac{\gamma}{k}\right), \quad q \neq -Q,$$

and the resulting bid-ask spread quoted by the market maker is given by:

$$\psi^*(t,q) = -\frac{1}{k}\ln\left(\frac{v_{q+1}(t)v_{q-1}(t)}{v_q(t)^2}\right) + \frac{2}{\gamma}\ln\left(1+\frac{\gamma}{k}\right), \quad |q| \neq Q.$$

Moreover, it is possible to consider an extension of this problem by adding market impact components:

$$dS(t) = \sigma dW(t) + \xi dN^a(t) - \xi dN^b(t), \quad \xi > 0. \tag{7.4}$$

The impact of MOs are characterized by constant ξ values, also refers to adverse selection risk of an MM. Equation 7.4 shows that whenever an LO is is filled on the bid size, the mid-price decreases and whenever an LO is filled on the ask side, the mid-price increases. Note that since the LOs are posted by the MM, they should be of same size. Actually, adverse selection describes the same interaction between the filled LOs and the change in reference price as well. Let us present the solution of the optimal control problem 7.3 with the market impact as follows:

Theorem 7.2

(Guéant et al. (2013) [18]) Let us consider a family of functions $(v_q)_{|q|\leq Q}$ *solution of the linear system of ODEs that follows:*

$$\forall q \in \{-Q+1,...,Q-1\}, \quad \dot{v}_q(t) = \alpha q^2 v_q(t) - \eta e^{-\frac{k}{2}\xi}(v_{q-1}(t) + v_{q+1}(t)),$$

$$\dot{v}_Q(t) = \alpha Q^2 v_Q(t) - \eta e^{-\frac{k}{2}\xi} v_{Q-1}(t),$$

$$\dot{v}_{-Q}(t) = \alpha Q^2 v_{-Q}(t) - \eta e^{-\frac{k}{2}\xi} v_{-Q+1}(t),$$

with $\forall q \in \{-Q,...,Q\}$, $v_q(T) = \exp(-\frac{1}{2}k\xi q^2)$, *where* $\alpha = \frac{k}{2}\gamma\sigma^2$ *and* $\eta = A(1+\frac{\gamma}{k})^{-(1+\frac{k}{\gamma})}$.

Then, $u(t,x,q,s) = -\exp(-\gamma(x+qs+\frac{1}{2}\xi q^2))v_q(t)^{-\frac{\gamma}{k}}$ *is the value function of the control problem.*

The optimal quotes are given by:

$$s - s^{b*}(t,q,s) = \delta^{b*}(t,q) = \frac{1}{k}\ln\left(\frac{v_q(t)}{v_{q+1}(t)}\right) + \frac{\xi}{2} + \frac{1}{\gamma}\ln\left(1+\frac{\gamma}{k}\right),$$

$$s^{a*}(t,q,s) - s = \delta^{a*}(t,q) = \frac{1}{k}\ln\left(\frac{v_q(t)}{v_{q-1}(t)}\right) + \frac{\xi}{2} + \frac{1}{\gamma}\ln\left(1+\frac{\gamma}{k}\right),$$

and the resulting bid-ask spread of the market maker is:

$$\psi^*(t,q) = -\frac{1}{k}\ln\left(\frac{v_{q+1}(t)v_{q-1}(t)}{v_q(t)^2}\right) + \xi + \frac{2}{\gamma}\ln\left(1+\frac{\gamma}{k}\right).$$

Moreover,

$$\lim_{T\to+\infty} \delta^{b*}(0,q) = \frac{1}{\gamma}\ln\left(1+\frac{\gamma}{k}\right) + \frac{\xi}{2} + \frac{1}{k}\ln\left(\frac{f_q^0}{f_{q+1}^0}\right),$$

$$\lim_{T\to+\infty} \delta^{a*}(0,q) = \frac{1}{\gamma}\ln\left(1+\frac{\gamma}{k}\right) + \frac{\xi}{2} + \frac{1}{k}\ln\left(\frac{f_q^0}{f_{q-1}^0}\right),$$

$$\lim_{T\to+\infty} \psi^*(0,q) = -\frac{1}{k}\ln\left(\frac{f_{q+1}^0 f_{q-1}^0}{(f_q^0)^2}\right) + \xi + \frac{2}{\gamma}\ln\left(1+\frac{\gamma}{k}\right),$$

where f^0 *is an eigenvector corresponding to the smallest eigenvalue of:*

$$\begin{bmatrix}
\alpha Q^2 & -\eta e^{-\frac{k}{2}\xi} & 0 & \cdots & \cdots & \cdots & 0 \\
-\eta e^{-\frac{k}{2}\xi} & \alpha(Q-1)^2 & -\eta e^{-\frac{k}{2}\xi} & 0 & \ddots & \ddots & \vdots \\
0 & \ddots & \ddots & \ddots & \ddots & \ddots & \vdots \\
\vdots & \ddots & \ddots & \ddots & \ddots & \ddots & \vdots \\
\vdots & \ddots & \ddots & \ddots & \ddots & \ddots & 0 \\
\vdots & \ddots & \ddots & 0 & -\eta e^{-\frac{k}{2}\xi} & \alpha(Q-1)^2 & -\eta e^{-\frac{k}{2}\xi} \\
0 & \cdots & \cdots & \cdots & 0 & -\eta e^{-\frac{k}{2}\xi} & \alpha Q^2
\end{bmatrix}$$

Furthermore, authors, Guéant et al. (2013) [18], develop the approximations for the bid-ask spread and the optimal quotes δ^{b*} and δ^{a*} as follows:

$$\delta_{\infty}^{b*}(q) \simeq \frac{1}{\gamma}\ln\left(1+\frac{\gamma}{k}\right)+\frac{\xi}{2}+\frac{2q+1}{2}e^{\frac{k}{4}\xi}\sqrt{\frac{\sigma^2\gamma}{2kA}\ln\left(1+\frac{\gamma}{k}\right)^{1+\frac{k}{\gamma}}},$$

$$\delta_{\infty}^{a*}(q) \simeq \frac{1}{\gamma}\ln\left(1+\frac{\gamma}{k}\right)+\frac{\xi}{2}-\frac{2q-1}{2}e^{\frac{k}{4}\xi}\sqrt{\frac{\sigma^2\gamma}{2kA}\ln\left(1+\frac{\gamma}{k}\right)^{1+\frac{k}{\gamma}}},$$

$$\psi^{*}(q) \simeq \frac{2}{\gamma}\ln\left(1+\frac{\gamma}{k}\right)+\xi+e^{\frac{k}{4}\xi}\sqrt{\frac{\sigma^2\gamma}{2kA}\ln\left(1+\frac{\gamma}{k}\right)^{1+\frac{k}{\gamma}}}.$$

Guéant et al. (2013) [18] provide several comparative results and illustrations describing the theorems proved in their work.

In the following section, we focus on the short term alpha and adverse selection problem of an MM by presenting an example studied in Cartea et al. (2015) [8].

7.4 A Focus: Short-term-alpha

In this section, we provide an approach to adverse selection risk by following the model developed in Cartea et al. (2015) [8], Section 10.4.2.

Firstly, let us present the intuition behind the model.

The price of a risky asset tends to move to its average price over time, that is: If the current market price of a stock is higher than the average price, then a fall in the price is expected. Hence, the investors intend to sell the asset. Meanwhile, if the market price is lower than the average price of the asset, as a consequence of the expectation of a price increase, the asset becomes attractive for purchase. This phenomenon is called *mean reversion* in finance. It can be used to analyze the trading range of the stock price and by using analytical techniques, the average price can be computed as well.

In this section, we assume that the dynamics of the stock price evolves as follows:

$$dS(t) = (\nu+\alpha(t))dt+\sigma dW(t),$$

where the drift component is driven by a long term component ν and a predictable, zero-mean reverting, short term process α. Observe that the short and long term components prevent MMs from unexpected loses against better informed traders. On the other hand, if an MM trades in a time scale in which she is able to observe α, then, she may post more speculative prices and escape from money losses.

Cartea et al. (2015) [8] focus on an α, which is driven by the order flow, and it is assumed that the MM is a high frequency trader.

Therefore, the dynamics of the short term component α can be defined by a jump-process, which jumps up, when a buy MO arrives and jumps down, when a sell MO arrives as described by the following dynamics:

$$d\alpha(t) = -\zeta\alpha(t)dt + \eta dW^{\alpha}(t) + \varepsilon^{+}_{1+M^{+}_{t-}} dM^{+}(t) - \varepsilon^{-}_{1+M^{-}_{t-}} dM^{-}(t),$$

where $\{\varepsilon_1^{\pm}, \varepsilon_2^{\pm}, ...\}$ are independent identical distributed random variables carrying the MO impact on the drift of the mid-price. Here, in this setting, ζ and η are positive constants and the intensities of the MOs are represented by $\lambda^{\pm}$.

The goal of the problem formulation is to count the influence of short-term α, when the MM posts only at the touch, i.e., at the best bid and ask.

Let $l^{\pm} \in \{0,1\}$ characterize whether the MM posts on the sell (+) side or on the buy (–) side of the LOB.

Thus, the cash process of the MM is defined as follows:

$$dX^{l}(t) = \left(S(t) + \frac{\Delta}{2}\right)dN^{+,l}(t) - \left(S(t) - \frac{\Delta}{2}\right)dN^{-,l}(t),$$

where the controlled doubly stochastic Poisson processes $N^{\pm,l}(t)$ count the number of MM's filled LOs with intensities $l^{\pm}(t)\lambda^{\pm}$, $t \in [0,T]$, and Δ denotes the constant spread.

Finally, the performance criteria for the MM is:

$$J^{l}(t,x,S,\alpha,q) = E_{t,x,S,\alpha,q}\left[X^{l}_{T} + Q^{l}_{T}(S_T - (\frac{\Delta}{2} + \varphi Q^{l}_{T})) - \phi\int_{t}^{T}(Q^{l}_{u})^{2}du\right],$$

where $\varphi \geq 0$ defines the fees for taking liquidity and the impact of an MO walking the LOB, besides, $\phi \geq 0$ describes the running inventory penalty parameter.

Authors, [8], assume that $\mathscr{A}$ denotes the set of $\mathscr{F}$-predictable strategies such that if the inventory is equal to upper (lower) inventory level $\overline{q}$ ($\underline{q}$), MM does not post a buy (sell) LO.

Finally, our goal is to find the optimal l and to demonstrate the value function V by solving the following optimal control problem:

$$V(t,x,S,\alpha,q) = \sup_{l\in\mathscr{A}} J^{l}(t,x,S,\alpha,q).$$

Consequently, the authors obtain the following HJB equations for the corresponding value function (see [8], Section 10.4.2):

$$
\begin{aligned}
0 = & \left(\partial_t + \alpha\partial_S + \frac{1}{2}\sigma^2\partial_{SS} - \zeta\alpha\partial_\alpha + \frac{1}{2}\eta^2\partial_{\alpha\alpha}\right)V - \phi q^2 \\
& + \lambda^+ \max_{l^+\in\{0,1\}} \left\{ 1_{q>\underline{q}}\, E\left[V(t, x+(S+\frac{\Delta}{2}l^+)l^+, S, \alpha+\varepsilon^+, q-l^+) - V\right] \right. \\
& \left. + \left(1 - l^+ 1_{q>\underline{q}}\right) E\left[V(t,x,S,\alpha+\varepsilon^+,q) - V\right] \right\} \\
& + \lambda^- \max_{l^-\in\{0,1\}} \left\{ 1_{q<\overline{q}}\, E\left[V(t, x-(S-\frac{\Delta}{2}l^-)l^-, S, \alpha-\varepsilon^-, q+l^-) - V\right] \right. \\
& \left. + \left(1 - l^- 1_{q<\overline{q}}\right) E\left[V(t,x,S,\alpha-\varepsilon^-,q+l^-) - V\right] \right\} \qquad (7.5)
\end{aligned}
$$

with respect to the terminal condition

$$
V(T,x,S,\alpha,q) = x + q(S - (\frac{\Delta}{2} + \varphi q)).
$$

Here, in Equation 7.5, the second and the fourth maximization terms characterize whether the MM posts an LO at-the-touch. The second line describes the change in value function, whenever the MM posts an LO and arrival of an MO lifts the offer and causes a jump in the short term alpha, simultaneously. The third line explains the change in value function whenever an MO arrives, however, the MM did not post an order. In this case, a jump occurs just in the dynamics of α. The fourth and fifth lines can be considered similarly, but on the bid side of the LOB. Therefore, an ansatz of the value function can be interpreted as follows:

$$
V(t,x,S,\alpha,q) = x + qS + h(t,\alpha,q). \qquad (7.6)
$$

By putting the function in Equation 7.6 into the HJB equations (7.5) and applying first order conditions, the optimal $l^{\pm}$ can be developed as in the following equations:

$$
\begin{aligned}
l^{+,*}(t,q) &= 1_{\left\{\frac{\Delta}{2} + E\left[h(t,\alpha+\varepsilon^+,q-1) - h(t,\alpha+\varepsilon^+,q)\right] > 0\right\} \cap \left\{q>\underline{q}\right\}}, \\
l^{-,*}(t,q) &= 1_{\left\{\frac{\Delta}{2} + E\left[h(t,\alpha-\varepsilon^-,q+1) - h(t,\alpha-\varepsilon^-,q)\right] > 0\right\} \cap \left\{q<\overline{q}\right\}}.
\end{aligned}
$$

Please see Cartea et al. (2015) [8] (Section 10.4.2) for numerical illustrations and for further details.

In the following section, we introduce some numerical approaches for market making in a European option market by addressing the results of Sağlam and Stoikov (2009) [32].

7.5 Market Making in an Options Market

Sağlam and Stoikov (2009) [32] assume that an MM trades just in a European call option with maturity T_{mat} and strike K. Furthermore, they delta-hedge their position at each time step. In this case, the authors focus on two types of risk: the Gamma risk, which arises due to an over night jump in the stock price, and the Vega risk, which is modeled directly via stochastic implied volatility.

The MM trades with LOs in the form of bid, $p^{b,o}$, and ask, $p^{a,o}$, quotes around the option's mid price, $C(t)$, as follows:

$$p^{b,o}(t) = C(t) - \delta^{b,o}(t),$$
$$p^{a,o}(t) = C(t) + \delta^{a,o}(t),$$

where $\delta^{b,o}$ and $\delta^{a,o}$ represent the spread (premium) gained by the MM at each trade.

The number of the options bought $N^{b,o}$ and sold $N^{a,o}$ before time t are represented by the Markov modulated Poisson processes with rates $\lambda(\delta^{b,o}(\cdot))$ and $\lambda(\delta^{a,o}(\cdot))$, respectively. By the way, the inventory in options is defined as follows:

$$q(t) = N^{b,o}(t) - N^{a,o}(t).$$

An appropriate choice for the intensity $\lambda^o(\delta)$ can be determined as a decreasing linear function:

$$\lambda^o(\delta) = \begin{cases} C - D\delta & \text{if } 0 \leq \delta < C/D, \\ 0 & \text{otherwise.} \end{cases}$$

Here, in this context, C and D can be adjusted to capture the relative liquidity of the option markets. Moreover, it is well known by empirical fact that the choice of a decreasing function with respect to λ justifies that if the quotes are closer to mid-price, the probability of their execution increases in a given time interval.

In [32], the authors decompose the cash process of MM into two parts: returns obtained by transactions $Z(\cdot)$ and the value of the inventory $I(\cdot)$.

Let us present the dynamics of $Z(\cdot)$:

$$dZ(t) = \delta^{a,o}(t)dN^{a,o}(t) + \delta^{b,o}(t)dN^{b,o}(t).$$

Now, let us introduce the market dynamics and define inventory process.

The trading day is divided into n sessions, $0 = t_0 < t_1 < t_2 < \ldots < t_n$. The MM sets the bid and ask quotes of a European option, $p_i^{b,o}$, $p_i^{a,o}$, respectively, and the delta of her option inventory is assumed equal to zero at the beginning of each session t_i for $0 \leq i \leq n-1$.

The mid-price of the stock is defined by the following continuous time dynamics:

$$dS(t) = \sigma(t)S(t)dW(t),$$

where the volatility σ is stochastic. Here, in this context, authors follow [30] to directly model the implied volatility of the option by:

$$d\hat{\sigma}(t) = \alpha dW^1(t),$$

where the Brownian motions W and W^1 are independent.

The mid-price of the European call option is given by the Black Scholes formula and $\Theta(t)$, $\Gamma(t)$, and $\Delta(t)$ are the standard Greeks as follows:

$$dC(S,t) = \Theta(t)dt + \Delta(t)dS(t) + \frac{1}{2}\Gamma(t)(dS(t))^2 + C_\sigma d\hat{\sigma}(t) + \frac{1}{2}C_{\sigma\sigma}(d\hat{\sigma}(t))^2.$$

By the approach of [30], the relation between stochastic volatility and implied volatility of the option is demonstrated by the following equation:

$$\sigma^2(t) = \hat{\sigma}^2(t) - \frac{\alpha^2}{\hat{\sigma}^2(t)}\left(\left[\ln\left(\frac{S}{K}\right)\right]^2 - \frac{1}{4}(T_{mat}-t)^2\hat{\sigma}^4(t)\right).$$

Moreover, this relation helps us escape the arbitrage opportunities and the change in the option value can be restated as:

$$\Delta C(t) = \Delta_i\sigma_i S_i u\sqrt{\Delta(t)} + \frac{1}{2}\Gamma_i\sigma_i^2 S_i^2(u^2-1)\Delta(t) + C_\sigma\alpha\eta\sqrt{\Delta(t)},$$

where u and η are standard normal random variables.

Note that by this equation, the risk of discrete hedging (the second term) and the risk of implied volatility (third term) can be caught.

Finally by considering the relation between Gamma and Vega, $C_\sigma = \Gamma S^2\sigma(T_{mat}-t)$, and keeping the assumption of delta-hedging of the options at each time step, we get

$$\Delta I_i = q_{i+1}^0\left(\frac{1}{2}\Gamma_i\sigma_i^2 S_i^2(u^2-1)\Delta(t) + \Gamma_i\sigma_i\alpha S_i^2(T_{mat}-t_i)\eta\sqrt{\Delta(t)}\right), \quad (7.7)$$

and

$$\Delta I_n = q_n^0\left(\frac{1}{2}\Gamma_n\sigma_n^2 S_n^2(u^2-1)\Delta(t) + \Gamma_n\sigma_n\alpha S_n^2(T_{mat}-t_n)\eta\sqrt{\Delta(t)}\right). \quad (7.8)$$

Remember that $C_n = C_{n-1}$, $S_n = S_{n-1}$, and $\sigma_n = \sigma_{n-1}$ are assumed, i.e., the prices of the option the stock, and the implied volatility will remain same in the last trading session. On the other hand, they may change in the overnight interval (t_n, T). Note that the inventory risk depends on higher order terms in Equations (7.7) and (7.8).

Consequently, our problem is to find the optimal spreads $\delta^{a,o}$ and $\delta^{b,o}$ in the options market by solving:

$$V(Z_{n-1}, S_{n-1}, q_{n-1}^s, q_{n-1}^o, t_{n-1}) = \max_{\delta_{n-1}^{a,o},\delta_{n-1}^{b,o}} E[Z_T|\mathscr{F}_{n-1}] - \gamma Var[\Delta I_n|\mathscr{F}_{n-1}].$$

Firstly, we present a theorem for a one-period model within the framework of [32].

Theorem 7.3
(Sağlam and Stoikov [32]) The optimal policy for the dealer is given by

$$\delta_{n-1}^{a,o} = \max\left(0, \min\left(\frac{C}{D}, \frac{C}{2D} - \gamma k\left(q_{n-1}^{o} - \frac{1}{2}\right)\right)\right),$$
$$\delta_{n-1}^{b,o} = \max\left(0, \min\left(\frac{C}{D}, \frac{C}{2D} + \gamma k\left(q_{n-1}^{o} + \frac{1}{2}\right)\right)\right),$$

where $k = (\frac{1}{2}\sigma^2(T - t_n) + \alpha^2(T_{mat} - t_n)^2)\Gamma_n^2 S_n^4 \sigma_n^2(T - t_n)$.

Now, let us provide the second theorem related to the multi-period model with no intraday movement as in [32]. Here, the problem is to solve:

$$V(Z_0, S_0, q_0^s, q_0^o, t_o) = \max_{\delta_i^{a,o}, \delta_i^{b,o}, 0 \le i \le n-1} E[Z_T] - \gamma Var[\Delta I_n].$$

and to find optimal bid and ask quotes of the option.

Theorem 7.4
(Sağlam and Stoikov (2009) [32]) The optimal bid and ask premiums at time t_i are given by

$$\delta_i^{a,o} = \max\left(0, \min\left(\frac{C}{D}, \frac{C}{2D} + m_{i+1}q_i^o - \frac{1}{2}m_{i+1}\right)\right),$$
$$\delta_i^{b,o} = \max\left(0, \min\left(\frac{C}{D}, \frac{C}{2D} - m_{i+1}q_i^o - \frac{1}{2}m_{i+1}\right)\right),$$

where m_i is the slope coefficient of the tilting and is calculated with the following recursion

$$m_i = m_{i+1} + \Delta(t)(Dm_{i+1}^2 J_i)$$

with the terminal condition

$$m_n = -\gamma\sigma_n^2 S^2 \Gamma_n^2 (T - t_n)(\frac{1}{2}\sigma_n^2 S^2 (T - t_n) + \alpha^2 S^2 (T_{mat} - t_n)^2)$$

and the auxiliary variable, $J_i = \left(1_{\{0 < \delta_i^{a,o} < \frac{C}{D}\}} + 1_{\{0 < \delta_i^{b,o} < \frac{C}{D}\}}\right)$.

We lead readers to Sağlam and Stoikov (2009) [32] for several results and descriptive graphs for market making in an options market.

7.6 Conclusion and Outlook

High Frequency Trading (*HFT*) is quantitative trading of short portfolio holding periods. The success and the profit of all portfolio-allocation decisions are largely driven by the strategies of high speed advanced computerized quantitative algorithms. HFT grew rapidly by its entry into the financial markets in the mid-2000s and now, for example, in the US equity market, it covers %50 of all trades. The main issue in trading via HFT is the high speed technologies. Actually, a high frequency trader holds the shares just for a fraction of a second before selling them. For example, let us assume that an HFT trader predicts that the price of a stock will rise just by a penny and just for a second, and then, it will drop back down. If the trader buys 1 million shares a moment before the price changes and sells them back after the increase, then she may profit thousands of dollars just in a second. This is how high frequency traders make a profit. Hence, it is not hard to guess that such management of this huge amount of information and such a microsecond, or millionths of a second trading activity can not be controlled by a human trader. Therefore, the whole process, not only executing the trade, but also searching for the signals of price movements, and moreover, making decisions about which trade to enforce are all done by specially designed super computers with direct access to exchanges. For a detailed analysis of HFT, see [2].

Market Making is a type of HFT. Consequently, all technological and regulative conditions are strongly related to MMs. In this context, the necessity of new investigations for algorithm developments such as *Reinforcement Learning* (*RL*) techniques are shining.

RL is a model-free approach in close relation to dynamic programming. It learns from experience without any knowledge of the underlying process and the goal is to maximize the cumulative reward (for a very detailed explanation of RL algorithms and techniques, see [34]). From the side of market making, the first attempt of applying this theory to solve the MM's, P&L problem is done by Chan and Shelton (2001) [11]. In this work, authors assume that however, there is no knowledge of the market environment, order arrival or price processes, etc., the agent learns via current market experience. Whereas authors developed a successful method, which converges on the expected strategies for uninformed traders, they did not focus on order cancellation or placement. Moreover, they skip the complexity of applying continuous state variables. Later, many authors pay attention to this hot topic in their very recent works [12, 14, 22, 24, 26, 31].

In this work, we provided several results obtained by traditional techniques of stochastic optimal control. Market making is a significant component of financial research and especially, in market microstructure. Therefore, it is clear that this theory will maintain its importance in a diversified path of scientific developments from theory to application. Obviously, there is still much work to be done.

Acknowledgement

This project is supported by SCROLLER: A Stochastic ContROL approach to machine Learning with applications to Environmental Risk models, Project 299897 from the Norwegian Research Council.

References

[1] F. Abergel, M. Anane, A. Chakraborti, A. Jedidi and I.M. Toke. *Limit Order Books*. Cambridge University Press, 2016.

[2] I. Aldridge. *High-frequency Trading: A Practical Guide to Algorithmic Strategies and Trading Systems*, volume 604. John Wiley & Sons, 2013.

[3] Y. Amihud and H. Mendelson. Dealership market: Market-making with inventory. *Journal of Financial Economics*, 8(1): 31–53, 1980.

[4] M. Avellaneda and S. Stoikov. High-frequency trading in a limit order book. *Quantitative Finance*, 8(3): 217–224, 2008.

[5] B. Biais, L. Glosten and C. Spatt. Market microstructure: A survey of microfoundations, empirical results, and policy implications. *Journal of Financial Markets*, 8(2): 217–264, 2005.

[6] Á. Cartea, R. Donnelly and S. Jaimungal. Algorithmic trading with model uncertainty. *SIAM Journal on Financial Mathematics*, 8(1): 635–671, 2017.

[7] Á. Cartea and S. Jaimungal. Risk metrics and fine tuning of high-frequency trading strategies. *Mathematical Finance*, 25(3): 576–611, 2015.

[8] Á. Cartea, S. Jaimungal and J. Penalva. *Algorithmic and High-frequency Trading*. Cambridge University Press, 2015.

[9] A. Cartea, S. Jaimungal and J. Ricci. Buy low, sell high: A high frequency trading perspective. *SIAM Journal on Financial Mathematics*, 5(1): 415–444, 2014.

[10] G. Chalamandaris and N.E. Vlachogiannakis. Adverse-selection considerations in the market-making of corporate bonds. *The European Journal of Finance*, 26(16): 1673–1702, 2020.

[11] N.T. Chan and C. Shelton. An electronic market-maker. Technical Report, Massachusetts Institute of Technology, Artificial Intelligence Laboratory, No: 2001–2005, 2001.

[12] H.L. Christensen, R.E. Turner, S.I. Hill and S.J. Godsill. Rebuilding the limit order book: Sequential bayesian inference on hidden states. *Quantitative Finance*, 13(11): 1779–1799, 2013.

[13] X. Gabaix, P. Gopikrishnan, V. Plerou and H.E. Stanley. Institutional investors and stock market volatility. *The Quarterly Journal of Economics*, 121(2): 461–504, 2006.

[14] S. Ganesh, N. Vadori, M. Xu, H. Zheng, P. Reddy and M. Veloso. Reinforcement learning for market making in a multi-agent dealer market. *arXiv preprint arXiv:1911.05892*, 2019.

[15] P. Gopikrishnan, V. Plerou, X. Gabaix and H.E. Stanley. Statistical properties of share volume traded in financial markets. *Physical Review E*, 62(4): R4493, 2000.

[16] M.D. Gould, M.A. Porter, S. Williams, M. McDonald, D.J. Fenn and S.D. Howison. Limit order books. *Quantitative Finance*, 13(11): 1709–1742, 2013.

[17] O. Guéant. Optimal market making. *Applied Mathematical Finance*, 24(2): 112–154, 2017.

[18] O. Guéant, C.-A. Lehalle and J. Fernandez-Tapia. Dealing with the inventory risk: A solution to the market making problem. *Mathematics and Financial Economics*, 7(4): 477–507, 2013.

[19] F. Guilbaud and H. Pham. Optimal high-frequency trading with limit and market orders. *Quantitative Finance*, 13(1): 79–94, 2013.

[20] T. Ho and H.R. Stoll. Optimal dealer pricing under transactions and return uncertainty. *Journal of Financial Economics*, 9(1): 47–73, 1981.

[21] T.S.Y. Ho and H.R. Stoll. The dynamics of dealer markets under competition. *The Journal of Finance*, 38(4): 1053–1074, 1983.

[22] M. Karpe, J. Fang, Z. Ma and C. Wang. Multi-agent reinforcement learning in a realistic limit order book market simulation. *arXiv preprint arXiv:2006.05574*, 2020.

[23] S. Laruelle, M. Rosenbaum and E.Savku. Assessing mifid II regulation on tick sizes: A transaction costs analysis viewpoint. *Market Microstructure and Liquidity*, 2020.

[24] I. Maeda, D. deGraw, M. Kitano, H. Matsushima, H. Sakaji, K. Izumi and A. Kato. Deep reinforcement learning in agent based financial market simulation. *Journal of Risk and Financial Management*, 13(4): 71, 2020.

[25] S. Maslov and M. Mills. Price fluctuations from the order book perspective—empirical facts and a simple model. *Physica A: Statistical Mechanics and its Applications*, 299(1-2): 234–246, 2001.

[26] J. Moody and M. Saffell. Learning to trade via direct reinforcement. *IEEE Transactions on Neural Networks*, 12(4): 875–889, 2001.

[27] M. O'hara. *Market Microstructure Theory*. Wiley, 1997.

[28] M. Potters and J.-P. Bouchaud. More statistical properties of order books and price impact. *Physica A: Statistical Mechanics and its Applications*, 324(1-2): 133–140, 2003.

[29] P. Sandås. Adverse selection and competitive market making: Empirical evidence from a limit order market. *The Review of Financial Studies*, 14(3): 705–734, 2001.

[30] P.J. Schönbucher. A market model for stochastic implied volatility. *Philosophical Transactions of the Royal Society of London. Series A: Mathematical, Physical and Engineering Sciences*, 357(1758): 2071–2092, 1999.

[31] T. Spooner, J. Fearnley, R. Savani and A. Koukorinis. Market making via reinforcement learning. *arXiv preprint arXiv:1804.04216*, 2018.

[32] S. Stoikov and M. Sağlam. Option market making under inventory risk. *Review of Derivatives Research*, 12(1): 55–79, 2009.

[33] R.J. Ritter. Investment banking and securities issuance. Constantinides, M. Harris and R Stulz (eds.) *Handbook of the Economics of Finance*. Amsterdam: North Holland, 2003.

[34] Sutton, Richard S. and Barto, Andrew G. *Introduction to Reinforcement Learning*, volume 135. MIT press Cambridge, 1998.

[35] P. Weber and B. Rosenow. Order book approach to price impact. *Quantitative Finance*, 5(4): 357–364, 2005.

Chapter 8

General Points of the Multi-criteria Flow Problems

Salima Nait Belkacem

8.1 Introduction

Operational research is the set of rational mathematical techniques permitting the elaboration of the best major component of decision aid, where the intellectual work becomes a simple execution work.

But when the luck and the competition appears daily life, the situation becomes more interesting. In fact, the English physician Patrick Blackett had the huge merit to lead the first heterogeneous operational researchers' crew in 1940; in which the works consist essentially in resolving military problems.

In 1896, multi-objective mathematical programming has been known for the first time by the Italian economist Pareto Vilfredo, but, it took until the 1950s–1970s for the problem with only one objective to be resolved. Otherwise, in many combinatorial optimization problems, the selection of the optimal solution takes more than one criterion. For example, for the transport or the flow network problems, the considered criterion could be the minimization of the cost, the necessary time, or the maximization of the security, etc.

However, these criteria are always in conflict, the reason why the formulation of the multi-objective network flow problem is a necessity. At the beginning of

University of M'hamed Bougara.

the 70s, the researchers became more interested in the bi-objective flow problems, this one is a particular case belonging to multi-objective linear programming.

In the real case, we see this kind of problem in the economic field, where we remark that the choice of the optimal action is based on the multiple and conflicting criteria, and to resolve this problem, we have to look for a compromise solution and make sure about its efficiency. It means that all these studies and researches done in the past by these scientists had a goal to help the decision maker to drive the network with better management of the system. So, we have to take into account the used criteria. Moreover, we should think about other objectives that may appear unexpectedly.

8.2 Basic Concepts of the Multi-Criteria Decision Aid

8.2.1 *Definitions of the set of actions and criteria*

Actions: The set of actions noted A, is the set of objects, decisions, candidates, solutions, etc., that we are going to explore in the decision process [1, 3, 23, 24, 25].

Criteria: We call criterion a function f defined on A, that takes its value in a totally ordered set, and that represents the decider's preferences according to a point of view.

Criteria coherence: A criteria coherent family is a family of criteria representing all the facets of the problem avoiding the repetitions.

8.2.2 *Definition of a multi-criteria problem*

A multi criteria decision problem is a situation where, having defined an action set A and a criteria coherent family F on A, we desire [14, 15, 19]:

- Either to determine an under-set of actions considered as the bests towards F (A choice problem).
- Or to partition A on under-sets following the pre-established norms (A selection problem).
- Or to put in order the actions of A, from the best one to the worst (arranging problem).

Dominance: Either two criteria victors $f^1, f^2 \in IR^r$, we say that f^1 dominates f^2 if and only if $f_j^1 \geq f_j^2$ for all j, and $f_j^1 > f_j^2$, at least for a value of $j \in \{1, ..., r\}$.

Efficient feasible solution: A feasible solution $x \in X$ of a network flow problem is efficient if and only if there is no other feasible solution $x' \in X$ as $f(x') \geq f(x)$ with $f(x') \neq f(x)$.

An efficient solution is also called non-dominated solution or Pareto optimal.

Efficient Edge: The edge connecting two adjacent efficient extreme points x^i and x^j is called an efficient edge.

Efficient point: Either $S \subset IR^n$, a point x of S is called node (or extreme point) if

$$\left.\begin{array}{l} x_1, x_2 \in S \\ x = 1/2x_1 + 1/2x_2 \in S \end{array}\right\} \rightarrow x = x_1, x_2.$$

Efficient base: Either B a base corresponding to an efficient extreme point x^i, and let x^j be an adjacent efficient extreme point, if we can obtain x^j from x^j by a pivot relation, then B is called efficient base.

8.3 Concepts of the Graph's Theory and Linear Programming

The spanning tree problem is the core of network optimization; it is used as a precious prototype in the combinatorial optimization that stimulates several supply lines [1, 2, 9, 20, 21].

8.3.1 Definitions

Graph: We call oriented graph and we note G = (V, A), a formed couple of:

set V of points or nodes generally finished.

A set $AV \otimes V$ of ordered pairs of nodes, or arcs.

Network: A network is a convex graph $R = (V, E)$ without buckles, consisting of a unique entry point S (source) and an exit point P (Shaft).

Tree: A tree is a connection graph without a cycle.

Spanning tree: A spanning tree is an oriented and connected graph without a cycle, possessing:

- Only one initial node, called root.
- Terminal nodes or leafs.
- Other descendants, having only one predecessor and at least one successor said knot.

Strong spanning tree A spanning tree is strong and feasible if the flow is sent positively from any node to the root of the tree, without changing the flow bound [1].

Arc: Every arc (xy) will be presented by an arrow of origin x and extremity y.

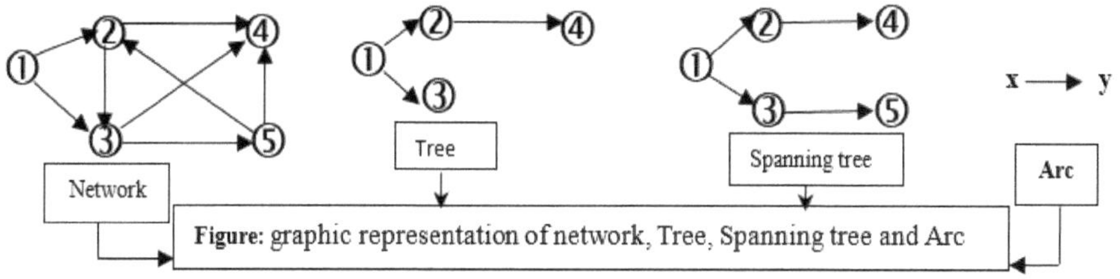

Figure 8.1: Regroups the different graphs notation.

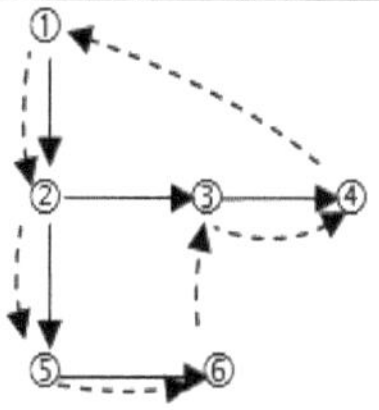

i	1	2	3	4	5	6
Pred(i)	0	1	2	3	2	5

This followed road signifies a knot : 1-2- 5- 6- 3- 4-1

Figure 8.2: Predecessor.

Base: A base is a set of $n-1$ arcs without a cycle. So it is the set of a tree arcs.

Predecessor: We say that y is a predecessor of x if there exists an arc having y as an initial extremity, and x as a terminal extremity.

Successor: We say that y is a successor of x, if there exists an arc having x as an initial extremity and y as a terminal extremity.

Chain: Two nodes x and y of a graph $G=(V,A)$ are on the same chain if z is another node of G, there exists a chain from x to z, and a chain from z to y.

Road: Either a graph $G=(V,A)$, we call road of G all continuation of nodes $x_0, x_1, \ldots, x_n$ as every couple (x_i, x_{i+1}) is an arc of A.

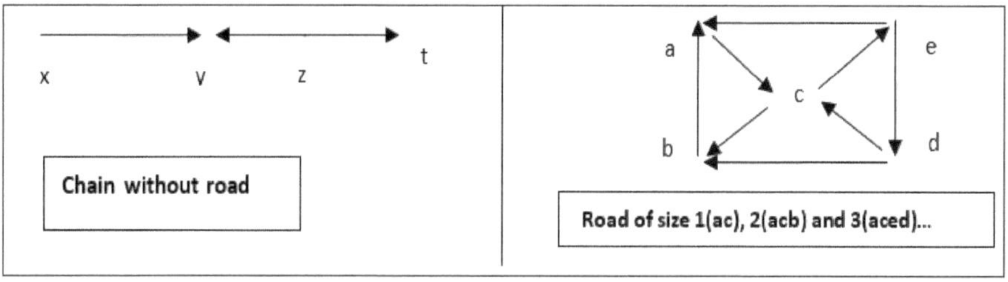

Figure 8.3: Graph representing circuit and cycle.

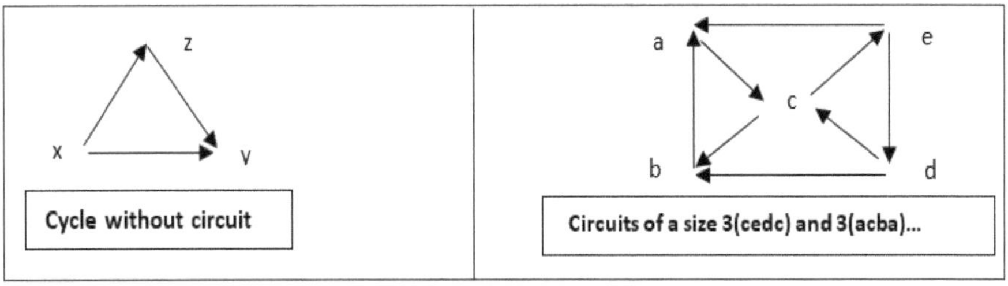

Figure 8.4: Graph representing chain and road.

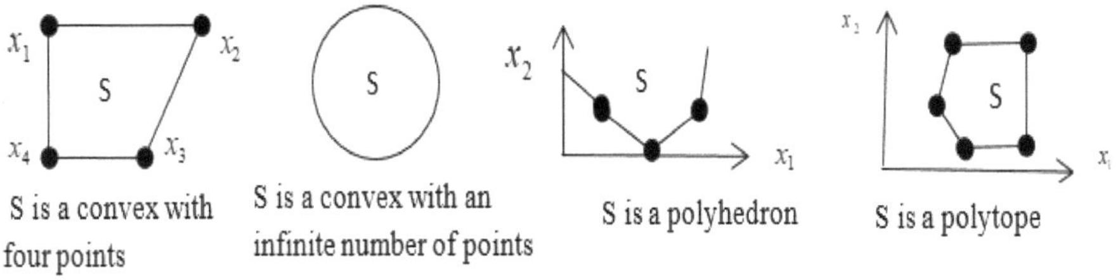

Figure 8.5: Different types of a convex and polyhedron.

Cycle: A cycle is a chain $(x_0, x_1, ..., x_n)$ in which the extremity coincides.

Circuit: Either a graph $G = (V,A)$, we call circuit of G, all road $(x_0, x_1, x_2, ..., x_{n-1}, x_0)$ in which the first node is also a final node.

Convex sets: A set $S \subset IR^n$ is convex if and only if, for all $x^1, x^2 \in S$, the point $(\lambda x^1 + (1-\lambda)x^2) \in S$ for a certain $\lambda \in [0,1]$.

Convex function: Either S a convex of IR^n, an application $f : S \longrightarrow S$ is called-convex if $\forall x, y \in S, \forall \lambda \in [0,1] f(\lambda x + (1-\lambda)y) \leq \lambda f(x) + (1-\lambda) f(y)$. A convex polyhedron is the intersection (eventually empty) of an infinite number of closed half-spaces. The different graphs of a convex and a polyhedron are illustrated in the Figure 8.5.

8.3.2 Hyperplans and half-spaces

Hyperplan: An hyperplan in IR^n of a dimension $(n-1)$ is the set $V = (\alpha \in IR$ and $c \in IR^n . c^T x = \alpha / c \neq 0)$ is intersection of two closed half-spaces. For example $V = \{(x,y) \in IR^2 / x + y = a\}$ is a hyperplan of a dimension 1.

Face: Either S a convex of IR^n, we call face of S, noted F all convex part of S as:

$$\left.\begin{array}{l} x \in F \\ x_1, x_1 \in F \\ x = \frac{1}{2}x_1 + \frac{1}{2}x_2 \end{array}\right\} \rightarrow x_1 \in F \text{ and } x_2 \in F.$$

As examples, $S = (3,6)$ is the intersection point of three half-spaces:

$\{(x_1,x_2) \in IR^2/2x_1 + x_2 \geq 12\}; \{(x_1,x_2) \in IR^2/x_2 \leq 6\}; \{(x_1,x_2) \in IR^2/x_2 \geq 0\}$(cone).

Alternatively, the hyperplan defined by $-x_1 + 3x_2 = 15$ has an intersection point with $S = (3,6)$.

Flow: A flow on a network $R = (V,E)$ with a capacity U and a values b_i may be defined as a function $F : E \to IR^+$, satisfying:
m constraints of capacity:

$$F(i,j) \leq U(i,j) \text{ for all } (i,j) \in E.$$

The flow x_{ij} that traverses the $arc(i,j)$ has to be inferior to the arc capacity.
n equations of a value b_i:

$$\sum_{(i,j)\in \pounds} x_{ij} - \sum_{(j,i)\in \pounds} x_{ji} = b_i \text{ for all } i \in V.$$

To assure the balance of the network, it is necessary to have $\sum_{j\in V} b_i = 0$it means: "verifying the Kirchhoff law".

The capacity $U : E \to R^+$ is associated to each arc (i,j), $U(i,j)$ is the maximal quantity of the flow, which transit from i towards j along this arc.

The cost $c : E \to R$ is associated to each arc (i,j), $c(i,j)$ is the passage cost from a flow unit along this arc.

The value $b_i : V \to R$ is associated to each node i, it is the difference between the flow leaving i and the one arriving on i. Here,

i) if $b_i > 0$, we send b_i flow unites on i (i is called producer node).

ii) if $b_i = 0$, there is a conservation of the flow on i (i is called transit node).

iii) if $b_i < 0$, we take out - b_i flow unites on i (i is called consumer node).

To this topological knowledges of the network, quantitative datum's will be added.

Lower bound of the flow: In certain problems we want to impose a minimum quantity of the flow transiting in the arcs, this lower bound is given by $L : E \to R$, then, the m capacity constraints becomes:

$$L(i,j) \leq f(i,j) \leq U(i,j) \forall \in E.$$

We model the arc (i,j) with the a lower bound $L(i,j) \neq 0$ decreasing the value b_i of the quantity $L(i,j)$, increasing the value b_i of $L(i,j)$ and fixing the capacity to $U(i,j) - L(i,j)$. The flow circulating on the modified arc, corresponds to the flow quantity in addition to the minimum quantity of $L(i,j)$.

Minimum cost flow problem: A minimum cost flow problem in a network $R = (V,E)$ with a capacity U, a cost c and a value b_i. It may be explained by the research of a flow x in a minimum cost. Otherwise, minimize the function $\sum_{(i,j\in)} c_{ij}x_{ij}$ comes to optimize the price that we have to pay to make the flow pass from the producer nodes to the consumer nodes through the network.

Pivot operation: Pivot operation is a movement from a spanning tree structure to another. The minimum cost flow problem treats other problems simultaneously.

Flow along cycles: Let w be a cycle (not necessarily directed) from a specified node s to itself, whose orientation has already been defined. Let $\bar{w}$ and $\bar{w}$ denote the sets of forward and backward arcs in this cycle, concerning the orientation of the cycle, we define the cycle multiplier $\mu(w)$ as follows [1]: $\mu(w) = \frac{\prod_{(i,j)\in\bar{w}} \mu_{ij}}{\prod_{(i,j)\in\underline{w}} \mu_{ij}}$ if we send 1 unit of the flow along the cycle w. Here,

i) If $\mu(w) > 1$, the cycle w as a gain cycle.

ii) If $\mu(w) < 1$, the cycle w as a loss cycle.

iii) If $\mu(w) = 1$, the flow around this cycle is conserves.

Maximum flow: To maximize the flow that passes in transit from a source node S to a Shaft node P through a network $R = (V,E,s,p)$ equipped with a capacity U, it is enough to give null arcs to the network and to add a return arc (P) with a cost $c(p,s) = -1$ and a capacity $U((p,s)) = \infty$, moreover, we don't consider any producer or consumer node.

Maximum flow at a minimum cost: To calculate a maximum flow between a source S and a shift P, minimizing the cost, we have to add a return $arc(P,S)$ with a capacity $U((p,s)) = \infty$ and a very negative cost. This problem may be easily formulated, either the network $R = (V,E)$ and incidence matrix $A \in R\ (V \times E)$ as following:

$$A_{v.(i,j)} = \begin{cases} 0 & v \neq j. \\ 1 & \forall v \in V \text{ and } (i,j) \in E. \\ -1 & v = j. \end{cases}$$

The variable of the problem $x \in R(E)$ characterises the function of the flow. The component $x_{(i,j)}$ (or x_{ij}) is equal to the flow on the $arc(i,j)$ and the minimum cost flow problem becomes:

$$\begin{cases} \text{minimize} c'x \\ \text{such that} \\ Ax = B \\ 0 \leq x \leq U \end{cases}$$

For example the graphic representation of network is shown in Figures 8.6, 8.7 and [3].

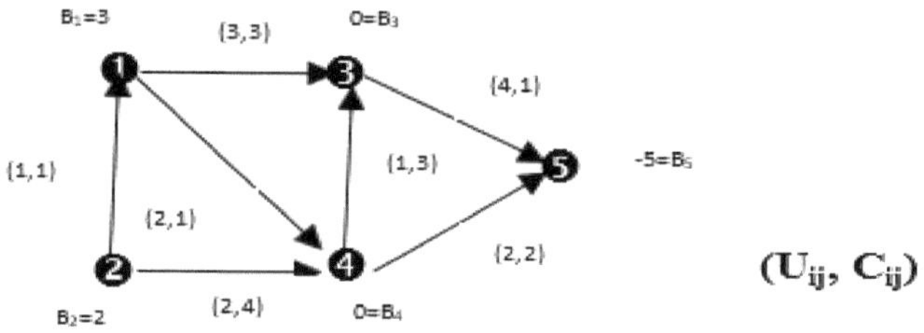

Figure 8.6: A network of 5 nodes and 7 arcs.

$Arcs(1,3)(1,4)(2,1)(2,4)(3,5)(4,3)(4,5).$

$c'x = (\,3 \quad 1 \quad 1 \quad 4 \quad 1 \quad 3 \quad 2).$

Sc

$$\begin{array}{l} \text{Nodes}: 1 \\ \quad\quad\quad\;\; 2 \\ \quad\quad\quad\;\; 3 \\ \quad\quad\quad\;\; 4 \\ \quad\quad\quad\;\; 5 \end{array} \mathbf{Ax} = \begin{Bmatrix} 1 & 1 & -1 & 0 & 0 & 0 & 0 \\ 0 & 0 & 1 & 1 & 0 & 0 & 0 \\ -1 & 0 & 0 & 0 & 1 & -1 & 0 \\ 0 & -1 & 0 & -1 & 0 & 1 & 1 \\ 0 & 0 & 0 & 0 & -1 & 0 & -1 \end{Bmatrix} \mathrm{x} = \begin{Bmatrix} 3 \\ 2 \\ 0 \\ 0 \\ -5 \end{Bmatrix} = \mathrm{B}$$

Figure 8.7: A network which can be represented by a linear program.

8.3.3 A linear program definition

A study begins by a modelling phase, a model is obtained in order to find a resolution algorithm, therefore, it is important to do an inventory of models equivalent to the flow problems.

Linear programming: Linear programming is an optimization branch, permitting the resolution of several problems. According to William J. Baumal, optimization is a mathematical technique (maximisation or minimisation) of a linear objective function under constraints, having the shape of a linear inequation or equation. It permits to select among the different actions, the one that is probably going to reach the aimed object. For R. Dorfman and P. Samuelson (1987) [8], linear programming is a determination method of the best action plan to realise objectives in a situation where the resources are limited. So, this programming is a resolution method of an economy problem, either in the field of a global economy or in a public sector [1, 2, 3, 12, 18].

A linear program can be formulated as follows:

$$(P1)\begin{cases} \text{Minimizer } \sum_{j=1}^{q} c_j x_j & \text{such as} \\ \sum_{j=1}^{q} a_{ij} x_j = b_i & \text{for each i=1,...,p.} \\ \mathrm{x}_j \geq 0 \text{ for each } j = 1, ..., q. & \end{cases}$$

This problem admits q decision variables x_j and p constraints.

The program (P1) is equivalent to the following (P2):

$$(P2)\begin{cases} \text{Minimize cx} \\ \text{Ax=b.} \\ \text{x} \geq 0. \end{cases}$$

In this expansion the matrix $A = (a_{ij})$ has p lines and q columns.

The vector $c = (c_j)$ is a line victor of dimension q.

The vector $x = (x_j)$ and $b = b_i$ are column victors of dimension q and p, respectively. Finally, $\Omega = x0,\ Axb$ is a polyhedron that represents a set of solutions.

On the other hand, a linear program of a minimum cost flow problem may be formulated by the following manner, where the network is noted by $R = (V, A, s, p)$ and A is a matrix:

$$(P3)\begin{cases} \text{minimize } \sum_{(i,j)\in A} c_{ij}x_{ij} & \text{as} \\ \sum_{j:(i,j)\in A} x_{ij} - \sum_{j:(i,j)\in A} x_j i = b_i & \text{for all i} \in V \\ \text{l}_{ij} \leq x_{ij} \leq u_{ij} & \text{for all (i, j)} \in A \\ \text{with } \sum_{i=1}^{n} b_i = 0 & \end{cases}$$

This program (P3) is equivalent to the following program (P4):

$$(P4)\begin{cases} \text{minimize} & \text{cx} \\ Ax = b & \text{x} \geq 0 \\ 1 \leq x \leq u & \end{cases}$$

where c_{ij} refers to cost by the flow unites associated to each arc, u_{ij} shows the maximum flow capacity. Moreover, l_{ij} is the Lower bound of the flow and x_{ij} represents decision variable that indicates the flow on the $arc(i, j)$, b_i denotes a relative integer number and finally, A presents a $(p \times q)$-dimensional incident matrix.

Fundamental theorem of a linear program: In a linear program if it admits a feasible solution, it admits a basic feasible solution, if it admits an optimal solution, it admits a basic optimal solution.

Lastly, if it admits a feasible solution and if the value of the objective function is narrow-minded (inferiorly if the objective function is to minimise and superiorly if the objective function is to maximize), it admits a basic optimal solution.

Theorem 8.1

Either a linear program

$(P)\begin{cases} Ax=b & x \geq 0 \\ cx=Z & max \end{cases}$ *. Here, a basic feasible solution of a linear program (P) corresponds to a convex polyhedron node of feasible solutions.*

Furthermore, a node of the convex polyhedron of the feasible solution corresponds to a feasible basic solution of the program.

8.4 Linear Programming Geometry

Geometric characterisation of the feasible solutions: The set S of feasible solutions of a linear programming problem is either

i) A narrow-minded polyhedron and not empty (polytope).

ii) A convex polyhedron not empty and not narrow-minded.

iii) An empty set.

Geometric characterisation of the optimal solutions: This can be grouped under three categories as follows:

i) If S is a polytope, either the optimal solution is unique and situated in a node of S, or there exist an infinity of optimal solutions that are the points of a side (face) of S, these solutions are then a convex combination of an infinite number of nodes.

ii) If S is a convex polyhedron not empty and not narrow-minded, it is impossible that the problem hasn't an optimal solution.

iii) If $S = \phi$, the problem has no feasible solution.

Remark: For a minimum cost flow network problem, each component of decision variable x corresponds to the arc flow, and the lines of matrix A corresponds to each network node.

8.5 Concepts of the Algorithmic Complexity

A similar problem might be resolved by several algorithms. That's why, it is important to compare their efficiency in which two principal criteria are generally considered. The first is the calculation time measuring the efficiency of the algorithm. The second is the necessary memory space at the execution of the algorithm. The main objective of the algorithm complexity is classifying every problem according to its difficulty.

The algorithmic complexity notions: An algorithm is a sequence of calculation steps permitting to pass from an entry value to an exit value, making the algorithm resolve a certain problem.

Its algorithmic complexity is an evaluation way of the cost, it measures the elementary operations number and the memory cost. This permits us to know the complexity of the problem in function of the instances size.

The complexity theory: The theory of the complexity was introduced to permit the measurement of the difficulty of an algorithmic problem in function of its instances size. Generally, the theory of complexity measures the difficulty of an

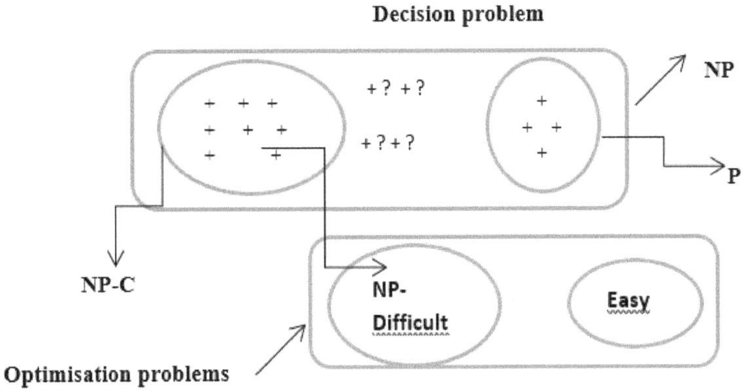

Figure 8.8: + problems, +? unclassified problems.

algorithm resolving the considered problem. There exist two principal types of problems: "Optimization problems ", where the objective is the minimization or the maximization of a certain property. And "Decision problems", asking for an answer of type "yes" or "no", see Figure 8.8 [6, 11, 4, 22].

8.5.1 The flow problems resolution method

Simplex method: The simplex method is an exact method to resolve the problems in a linear program. Its algorithm is the most popular, because it resolves the optimization problems with an economic gain. The mechanism of this method is as follows: Each transit from a node to another is represented by an iteration. The act of passing by all the nodes of the set gives us a (convex polyhedron). Starting from one of these nodes, the algorithm determines an edge of the allowable set (face of dimension) along which the cost function decreases and takes a new iteration, the node situated at the end of this selected edge, and that's what we call a "pivot operation" [1, 12, 10, 17]. As an example, consider the linear program P below,

$$(P)\begin{cases} \text{Maximize } Z = x_1 + x_2, \\ 2x_1 + 3x_2 \leq 12, \\ x_1 \leq 4, \\ x_2 \leq 3, \\ x_1, x_2 \geq 0. \end{cases}$$

In this example, the simplex method uses the extreme point property. It begins from a feasible extreme point and visits the adjacent extreme points, improving the solution value of the objective function from a stage to another until it reaches the optimal extreme point. It means that if the simplex method starts at point A, it has to visit the point B and C before reaching the point D, or it has to follow the road $A - E$ and D. See Figure 8.9.

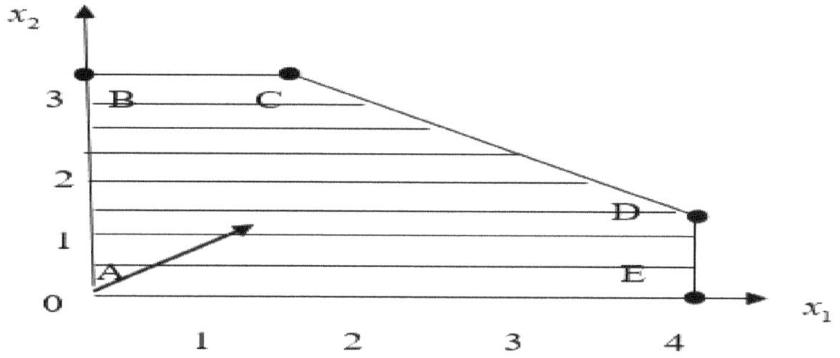

Figure 8.9: A set of feasible solutions for a linear program.

Simplex network algorithm: The central concept of the simplex network algorithm is the spanning tree notion. This algorithm is a movement from a spanning tree structure to another until reaching the optimal spanning tree structure. This movement permits us to complete all the operations on the same network without maintain the simplex table. Otherwise, it maintains the feasible spanning tree structure at each iteration and transforms it successively until it becomes optimal, and this is done at each stage by an arc changing [1, 2, 7, 13].

8.6 Conclusion

The application of the decision aid methods on the different projects is a complicated exercise that necessitates a compromise between many appearances. The profit cost analysis is often used as a decision gadget. But, the issue of these projects calls for a complicated set of criteria, such as the environmental impact, the security, the risks, etc. [5, 4, 19, 16].

These are difficult to convert into monetary values. So, the multi criteria analysis' methods are accompanied by its graph notions and formulations. It seems the most suitable choice for the decision aid under these kind of problems. This is done in order to choose the best action to implement at the end of the resolution stages of each project, taking into account the risks, time, the cost to minimize and the security as well as the quality to maximize, etc. These decision aid tools help the decision maker to formulate a clearer, a fairer and above all, a beneficial choice. These techniques should not be considered as definitive answers that aid decision; they are a part of the decision making.

Finally, if we want to succeed in any project, the best way is to apply decision aids. It is the pillar of a perfect company, and sometimes the key of the success of several projects.

References

[1] R.K. Ahuja, T.L Magnanti, J.B. Orlin and M.R. Reddy. Applications of network optimization. *Handbooks in Operations Research and Management Science*, 7: 1–83, 1995.

[2] K.A. Andersen, K. Jörnsten and M. Lind. On bicriterion minimal spanning trees: An approximation. *Computers & Operations Research*, 23(12): 1171–1182, 1996.

[3] S. Ben-Mena. Introduction aux méthodes multicritères d'aides *à* la decision. *Biotechnologie, Agronomie, Société et Environnement*, 2000.

[4] C. Bennadji. Marchés publics et corruption en algérie. *NAQD*, 1: 135–153, 2008.

[5] H.I. Calvete and P.M. Mateo. An approach for the network flow problem with multiple objectives. *Computers & Operations Research*, 22(9): 971–983, 1995.

[6] S.A. Cook. The complexity of theorem-proving procedures. In *Proceedings of the Third Annual ACM Symposium on Theory of Computing*, pp. 151–158, 1971.

[7] W.H. Cunningham. A network simplex method. *Mathematical Programming*, 11(1):105–116, 1976.

[8] R. Dorfman, P.A. Samuelson and R.M. Solow. *Linear Programming and Economic Analysis*. Courier Corporation, 1987.

[9] A. Eusébio, J.R. Figueira and M. Ehrgott. On finding representative non-dominated points for bi-objective integer network flow problems. *Computers & Operations Research*, 48: 1–10, 2014.

[10] R. Faure, B. Lemaire and C. Picouleau. *Précis de recherche Opérationnelle-7e éd.: Méthodes et exercices d'application*. Dunod, 2014.

[11] M.R. Garey and D.S. Johnson. *Computer and Intractability: A Guide to the Theory of NP-Completeness*, W.H. Freedman and Company, 1979.

[12] S. Hadj-Messaoud. *Propriétés magnétostatiques et résonance ferromagnétique de réseaux de nanofils en configuration multicouche*. Ecole Polytechnique, Montreal (Canada), 2012.

[13] H.W. Hamacher, C.R. Pedersen and S. Ruzika. Multiple objective minimum cost flow problems: A review. *European Journal of Operational Research*, 176(3): 1404–1422, 2007.

[14] D. Klingman and J. Mote. Solution approaches for network flow problems with multiple criteria. Technical report, Texas Univ At Austin Center For Cybernetic Studies, 1979.

[15] H. Lee and P.S. Pulat. Bicriteria network flow problems: Integer case. *European Journal of Operational Research*, 66(1): 148–157, 1993.

[16] G. Nicollet and G. Labadie. Modèles hydrauliques fluviaux. Électricitéde France, *Direction desétudes et recherches*, Service, 1993.

[17] V. Pareto. Cours d'economie politique. bousquet, gh, g. busino, eds. *Oevres Completes de Vilfredo Pareto*, 1, 1896.

[18] J. Pictet and D. Bollinger. Aide multicritère à la décision. Aspects mathématiques du droit des marchés publics. *Baurecht/Droit de la Construction*, 2: 63–65, 2000.

[19] P.S. Pulat, F. Huarng and H. Lee. Efficient solutions for the bicriteria network flow problem. *Computers & Operations Research*, 19(7): 649–655, 1992.

[20] M. Sakarovitch. *Optimisation combinatoire: Programmation discrete, ser. collection enseignement des sciences*. Hermann, Volume 2, 1984.

[21] A. Sedeño-Noda and C. González-Martın. The biobjective minimum cost flow problem. *European Journal of Operational Research*, 124(3): 591–600, 2000.

[22] R. Sedgewick and P. Flajolef. *An Introduction to the Analysis of Algorithms*. Addison-Wesley Pearson Education. 2013.

[23] R.E. Steuer. Multiple criteria optimization. *Theory, Computation and Applications*, 1986.

[24] J. Teghem and P.L. Kunsch. A survey of techniques for finding efficient solutions to multi-objective integer linear programming. *Asia-Pacific Journal of Operational Research*, 3(2): 95–108, 1986.

[25] P. Vincke. *La Modélisation des Préférences*. PhD thesis, Institute de Mathématiques Économiques (IME), 1985.

Chapter 9

Operation Research in Neuroscience: A Recent Perspective of Operation Research Application in Finance

Betül Kalaycı,[1] *Vilda Purutçuoğlu*[1,*] and *Gerhard Wilhelm Weber*[2]

9.1 Introduction

Financial decisions are affected according to human neurologic activities. What is happening in investors' brains while making financial decisions such as selling and buying? Which hormones are activated when they are in a risky or risk-averse situation? are some examples of interesting questions on the human side. Hence, the studies on these topics are explained in a field called *neurofinance* which examines the neurological basis of the mental state in financial decisions. Neuroscience aims at connecting finance with the interdisciplinary area of Neurofinance, that comprises all the phenomena and relations, explaining feelings and thoughts that escape from our comprehension. Investors' financial performances emotionally reactive and lead to poor impulse control. They can be calculated by

[1] Middle East Technical University.
[2] Poznan University of Technology.
Email: betulkalayci@gmail.com
* Corresponding author: vpurutcu@metu.edu.tr

using numerous neuroscience techniques. These techniques provide a determination of the regions in the brain that are quite active in the course of decision making processes and, especially, decision making in finance [28, 30]. On the other hand, economists study several empirical facts regarding the behavior of individual investors apart from their hormonal or brain activities, such as how their emotions, opinions, and views influence their decisions. This constitutes the new subject of *Behavioral Finance* [8, 13, 15]. The general name of all these emotional states is called *Sentiment*. Behavioral Finance investigates the effects of Investor Sentiment in stock markets and in the economy. This is reflected in dealing with beliefs about the world as expectations about variables' conditions upon current data or current information [13, 15].

These kind of human factor and financial data might be subject to high fluctuation and high correlation, the traditional statistical methods usually are not able to cope with these complex problems [31]. Because of this high possibility of fluctuation, these traditional models have brought about a growing interest in machine learning techniques [3, 16]. Many investors' decisions and opinions in the finance industry are studied by machine learning techniques. This helps the finance industry comprehend the data and achieve a competitive advantage from it [14]. In addition, recently, hybrid models that combine several machine learning techniques tend to improve gradually [14, 9]. The association derived by machine learning has proved successful for hybrid methods. However, the process should always take into consideration the significance of data quality, as the ambiguity of data always requires the robustness of the machine learning technology [14].

In this study, we concentrate on the behavior of financial problems which is based on the investors' behavior introduced as Sentiment. Besides, we also compare the forecasting performance of the sentiment index by using the single MARS, RF, NN models, and MARS-NN, MARS-RF, RF-MARS, RF-NN, NN-MARS, and NN-RF hybrid models. Here, MARS denotes Multivariate Adaptive Regression Splines, NN implies the Neural Network model and RF indicates the Random Forest approach.

As mentioned above, since our study is based on forecasting and the optimization problem, we discourse operations research (OR) is a fundamental tool for decision theory with uncertainty. This analytical method of "problem-solving" and "decision-making" uses quantitative methods. The benefits of using methods of OR in many applications and machine learning fields is evident.

Recently, there are some studies that show the relation of machine learning techniques in Operational Research (OR). Talavera and Luna (2020) present how machine learning techniques such as NN, Fuzzy logic, etc., are integrated in applications in an Operation Research course. To optimally come up against real applications in decision-making problems, they united computational intelligence techniques into the course of Operation Research [29].

In the following sections; firstly, the machine learning methods which we use in this study are introduced. Secondly, we give information about investors' behavior and the term "Sentiment". Because, in the literature, there are a lot of studies for machine learning techniques applied to finance. In this part, we briefly mention those and afterward we examine the application and discussion of the results of the application.

9.2 Machine Learning Techniques

9.2.1 *Multivariate adaptive regression splines (MARS)*

MARS can be seen as a form of stepwise linear regression [11]. Unlike additive models, MARS does not treat the input variables separately and takes into account their interactions.

MARS is described as an approach of using smoothing splines to fit the relationship between a finite number of regressors and a response variable. In a piecewise sense, it yields a very smooth line, or surface that can capture "shifts" in the relationship between these variables. These shifts occur at positions assigned as "knots" and ensure a smooth transition between "regimes". The model does not only investigates all possible knot positions, but also, across all variables and all mutually affected positions among them. This is carried out via the use of integration of variables named as *basis functions*, called splines. When the optimal numbers of basis functions and knot positions are designated by MARS, the estimates of the fitted value via the chosen basis functions are provided with a final least-squares regression [27]. As a consequence, the resulting multivariate additive model is determined with a two-stage process, which is called the *forward stage* and the *backward stage*.

In the *forward stage*, MARS finds which basis functions (BFs) are attached to the model by using a fast searching algorithm, and builds a large model that usually overfits the dataset. The process continues until the model reaches the maximum number of basis functions, that is a specific value set by the users. In fact, BFs contribute both most and least to the overall performance together in this model. That is why the model is more complicated and contains many inaccurate terms in the forward stage. In the *backward stage*, the overfit model is clipped to diminish the complexity of the model. However, the model supports the overall performance taking into consideration the fit with the data. In this backward stage, BFs which contribute to the smallest raise in the *residual sum of squares (RSS)* are eliminated from the model at each iteration. Consequently, an optimally estimated model is created [11, 19, 23, 24].

MARS uses extansions in these piecewise linear one-dimensional basis functions of the form $(x-t)_+$ and $(x-t)_-$:

$$(x-t)_+ = \begin{cases} x-t, & \text{if } \; x>t, \\ 0, & \text{otherwise}, \end{cases} \quad \text{and} \quad (x-t)_- = \begin{cases} t-x, & \text{if } \; x<t, \\ 0, & \text{otherwise}. \end{cases}$$

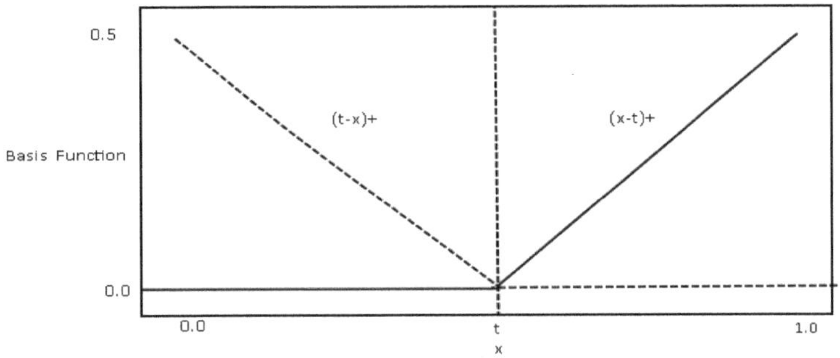

Figure 9.1: The BFs used by MARS for $t = 0.5$.

Each such a map is a *truncated linear function*, with a univariate knot at value t, determined using the dataset. Figure 2.1 shows the BF pairs for $t = 0.5$ as an example.

These two functions are called a *reflected pair*. The aim is to present a reflected pair for each input X_j with knots at each observed values x_{ij} of that input. Thus, the aggregation of the basis functions is

$$C = \{(x_j - t)_+, (x_j - t)_- : \quad t \in \{x_{1j}, x_{2j}, ..., x_{Nj}\}, j = 1, 2, ..., p\}, \tag{9.1}$$

where N is the number of the observations, p represents the dimension of the input space so that, in the case all of the input values are distinct, there are $2Np$ one-dimensional basis functions. Although each of the aforementioned basis functions depend only on a single X_j, such as by $h(x) = (x_j - t)_+$, it is regarded as a function over the whole input space $\mathbb{R}^p$.

As stated earlier, the "model-building strategy" begins like a "forward stepwise linear regression", but unlike using the original inputs, it is allowed to involve functions from the set C and their products. As a result, the model has the first following form:

$$f(x) = \beta_0 + \sum_{m=1}^{M} \beta_m h_m(x). \tag{9.2}$$

This was in the forward stage of MARS, where $\mathcal{M}$ is the set of basis functions in the present model, and $h_m(x)$ are multivariate basis functions from C or products of two or more such functions. Furthermore, β_m stands for the unknown coefficients at the constant 1 $(m = 0)$ or the mth basis function, and $x = (x_1, x_2, ..., x_p)^T$. The form of the mth basis function can be shown as follows:

$$h_m(x) = \prod_{k=1}^{K_m} \left(s_{km} \cdot (x_{v(k,m)} - t_{km})\right)_+ \tag{9.3}$$

in which K_m is the number of "truncated linear functions" multiplied in the mth basis function and $x_{v(k,m)}$ is the input variable which refers to the kth truncated linear function in the mth basis function. Furthermore, t_{km} denotes the knot value suitable to the variable $x_{v(k,m)}$ and $s_{km} = \pm 1$.

Herein, a *lack-of-fit criterion* is performed to assess the probable basis functions.

To form the model, the MARS forward stepwise algorithm begins with the constant function $h_m(x) = 1$ to estimate β_0, and the entire elements in the set C are candidate functions. There are some obtainable forms of the basis functions $h_m(x)$ which are listed as below.

- 1,
- x_k,
- $(x_k - t_i)$,
- $x_k x_l$,
- $(x_k - t_i)_+ x_l$,
- $(x_k - t_i)_+ (x_l - t_j)_+$.

For each of these basis functions, the input variables cannot be identical in the MARS algorithm. Therefore, the basis functions which are stated above use different input variables, x_k and x_l, and their knots t_i and t_j. We take into account as a new basis function pair for all products of a function h_m from the model set $\mathcal{M}$ with one of the reflected pairs in C, at each stage. For the model set the term of the form is added which generates the biggest decline in the training error [11, 23, 24]:

$$\hat{\beta}_{M+1} h_l(x) \cdot (x_j - t)_+ + \hat{\beta}_{M+2} h_l(x) \cdot (t - x_j)_+, \qquad h_l \in \mathcal{M}. \qquad (9.4)$$

In the expression 9.4, $\hat{\beta}_{M+1}$ and $\hat{\beta}_{M+2}$ are coefficients determined by the least square (LS) estimation, together with the rest of the $M+1$ coefficients in the model. Then, the resulting products are supplemented by the model, and the process ends when the model set $\mathcal{M}$ includes a maximum number of predetermined terms. To give an instance, the subsequent basis functions can be possible candidates of being added to the model [11, 24]:

- 1,
- x_k,
- $(x_k - t_i)$, if x_k is already in the model,
- $x_k x_l$, if x_k and x_l are already in the model,
- $(x_k - t_i)_+ x_l$, if $x_k x_l$ and $(x_k - t_i)$ are already in the model,
- $(x_k - t_i)_+ (x_l - t_j)_+$ if $(x_k - t_i)_+ x_l$ and $(x_k - t_i)_+ x_k$ are already in the model.

In the final of this forward process at each step, we have obtained a large model which frequently overfits the data. Therefore, a backward deletion is applied, too. At each step, we remove the term which leads to the smallest rise in the residual squared error from the model by using a backward algorithm. This procedure stops when an optimal number of effective terms is obtained in the last model. Then, this produces an estimated best model $\hat{f}_\lambda$ of any size (i.e., number of terms) λ. In the MARS model, the so-called *Generalized Cross-Validation (GCV)* is employed to estimate the optimal value of λ. This criterion is identified as

$$GCV(\lambda) = \frac{\sum_{i=1}^{N}(y_i - \hat{f}_\lambda(x_i))^2}{(1 - M(\lambda)/N)^2}. \tag{9.5}$$

In Equation 9.5, the value $M(\lambda)$ is the influential number of parameters in the model and N is the number of sample observations.

Some classical statistical methods such as regression techniques can be good at conducting interaction terms. These methods need to try many combinations of the variables in the dataset. Hence, they cannot provide a computationally efficient solution. On the other hand, MARS automatically searches for appropriate interactions among independent variables, which can be particularly preferred whenever there is a large number of interactive variables. As a result, it can identify interactions and a comparatively small number of regressor variables which are complex transformations of initial variables. In addition to these advantages, it also provides an opportunity to discover nonlinearities that can appear in the relationship between dependent and independent variables while producing graphs that help visualize and understand the interactions [11, 24].

9.2.2 Random forest algorithm

With more mathematical details, the random forest algorithm is one of the supervised learning methods in the field of the data mining techniques. In this algorithm, classification and regression are considered. In classification, the goal is to assign each observation to the correct subgroups whose elements are previously known [7, 26].

Mathematical Details of the Random Forest Algorithm

The random forest algorithm (RF) selects the most voted class among the classes consisting of many generated trees and the outputs of the class that is the mode of the class output by individual trees [26, 31]. This algorithm is also explained as a combination of the tree-structured classifiers in the forest. The random forest has superiority on bagging as it builds each tree as regards different bootstrap samples and it carries out its classification as the tree construction method. In opposition to the standard trees, in RF, small subsets of predictors are generated and under the random selection, each node is split by employing the best one among these randomly selected predictors. In RFA, as the forest becomes comprising more trees, an upper boundary for the generalization error $(PE*)$ is created to avoid the overfitting problem without requiring large datasets. Therefore, we do not encounter such a problem in the RF algorithm (RFA) while other algorithms may result in an overfitting challenge. The main rule in RFA is to maximize the strength between nodes having the smallest correlation, the adaptive bagging is a helpful basis because it a good estimate of the main measurements of strength, correlation and generalization error with an additional process of limiting the generalization error by creating an upper boundary raise the accuracy of trees, motifs, modules and the resulting networks. The strength of each individual tree and the correlation between combinations of these trees determine the generalization error of a random forest. Moreover, the employ of the random feature selection to split each node results in an error rate comparable to the others. In addition to this, the generalization error, strength and the correlation can be seen by the internal estimates, and these estimates are used to present the response to the raising of the number of features involved in the splitting step. The internal estimates can also be applied to measure the importance of the variable. Moreover, in random forests, small communities (ensembles) are created and these ensembles vote for the most popular class. To generate these small communities, the most well-known way is to create random vectors denoted by θ. To refer the kth tree in the forest, the random vector representing that specific tree is taken as θ_k. In a forest, random vectors are $\theta_1, \theta_2, ..., \theta_{k-1}$ resulting in θ_k and a classifier $h(X, \theta_k)$, where X is an input vector and k also represents the convergence of trees in the forest. The growth of each tree in the small communities is checked by these vectors. Bagging, selecting the random split, and selecting the training set from a random set of weights are some ways to generate these random vectors. The accuracy of a random forest can be defined in terms of the generalization error, the strength of the individual tree classifiers, and the dependence measure which is the correlation between these classifiers. Thus, the generalization error (PE*) of RFA is checked as shown in Equation 9.6.

$$\text{PE}* \leqslant \frac{\bar{\rho}(1-s^2)}{s^2}. \tag{9.6}$$

In this inequality, the mean value of the correlation between random vectors θ and $\theta^{'}$ is indicated by $\bar{\rho}$ while is $\theta^{'}$ the proposal tree in the next iteration. Here, s denotes the strength of the small communities via $s = E_{X,Y}mr(X,Y)$. E defines the expectation between the random vectors X and Y, while the margin function is shown by $mr(.)$. There exist some key points in the random forest algorithm that need to be known how they are created to build the trees accurately. For instance, the strength and the correlation are also as important as the generalization error in the random forest algorithm. Creating an ensemble of classifiers, the representation of this committee is taken as $h_1(x), h_2(x), ..., h_k(x)$. By using these given properties, the margin function can be defined as below. 9.7:

$$\text{mg}(X,Y) = \text{av}_k I(h_k(X) = Y) - max_{j \neq Y}\text{av}_k I(h_k(X) = j). \tag{9.7}$$

In Equation 9.7, Y and X are the random vectors $(Y = [1...J]).I(h_k(X) = ..)$ denotes the indicator function and av_k is the average of the kth tree.

On the other side, the upper boundary of the generalization error has been mentioned previously. In order to elaborate that description, now, we define how it is determined by the margin function. By extending it, the generalization error represented by the margin function can be shown as in Equation 9.8.

$$\text{PE}* = P_{X,Y}(\text{mg}(X,Y) < 0). \tag{9.8}$$

In the above expression, the generalization error is represented by PE*.

As the rule of the random forest algorithm, the number of trees should be increased. In this situation, for all the ensemble $(\theta_{z1},...)$, the generalization error converges to the formula as shown in Equation 9.9.

$$P_{X,Y}(P_\theta(h(X,\theta) = Y) - max_{j \neq Y} P_\theta(h(X,\theta) = j) < 0), \tag{9.9}$$

where $max_{j \neq Y} P_\theta(h(X,\theta) = j)$ denotes the maximum of all the probability values among all the values of the classifier except for its value on the point Y and $h(X,\theta)$ denotes the classifier for the random vector X.

Since the rule in the random forest is to maximize the strength and to minimize the correlation, the strength is one of the most important features in the random forest algorithm for the process of the tree construction. Thus, by looking at the margin function, it is possible to define the strength of the ensemble of classifiers by Equation 9.10.

$$s = E_{X,Y} mr(X,Y). \tag{9.10}$$

Now it is appropriate to connect the generalization error and the strength as in following representation:

$$\text{PE}* \leqslant var(mr)/s^2. \tag{9.11}$$

In Equation 9.11, PE* is the generalization error as desribed before and s^2 refers to the square of the strength. By inserting the correlation into these equations, we see the following represantation.

$$V(mr) = \bar{\rho}(E_\theta sd(\theta))^2 \leqslant \bar{\rho} E_\theta V(\theta). \tag{9.12}$$

In Equation 9.12, $\bar{\rho}$ refers to the mean value of the correlation between θ and θ', E presents the expectation, sd denotes the standard deviation and $V(.)$ implies the variance.

Thus, for the definition of an upper boundary for the generalization error, all the above formulas are combined as shown in Equation 9.6.

The internal estimates are calculated to decide how many features will be chosen in each node. These internal estimates belong to the generalization error, the classifier strength and the dependence, which are also called as the *out-of-bag* (OOB) estimates.

The numerous uses of out-of-bag estimates are listed as follows [26]:

- OOB estimates are applied as an ingredient in the estimates of the generalization error.
- OOB estimates of the variance are used to infer the generalization error for the arbitrary classifiers.
- It is proved that the OOB estimate is accurate for using a test set of the same size as the training set.
- By using the OOB error, the estimate removes the need for a set aside the test set.

9.2.3 Neural network

Neural network is a model of the simulation of the human nervous system. The human nervous system is consists of cells, which will be mentioned as neurons [1]. These neurons are units of computation. They get input from other neurons, compute these inputs, and feed them into other neurons. The computing function which is placed at a neuron is identified by the weights on the input links to that neuron. This weight can be seen as analogous to the strength of a synaptic link. The computation function can be obtained by switching these weights conveniently, which is analogous to the learning and obtaining of the synaptic strength in biological neural networks. The "external stimulus" in artificial neural networks for finding out these weights is ensured by the training data. Hereby, the main goal is to gradually alter the weights whenever inaccurate predictions are made by the current set of weights [1].

Artificial neural network (ANN) is a matchless statistical technique that involves extreme computing power, powerful memory, learning ability, and error tolerance ability. This means ANN is a computing system that likens to the interlinking of neurons in organisms for complicated information processing. Neural network is an adaptive system which has the ability to learn. ANN can be trained to ensure requested output via varied algorithms [19].

Highly interconnected, interacting processing units that are propped up neuro-biological models constitute a *Neural Network* system. Neural networks process information through the interactions of a wide range of processor and their links to extrinsic inputs [19]. The impact of the neural network is the arrangement of the links among nodes. Here, Neural Network is an architecture that is responsible for this arrangement. There are a wide range of architectures, beginning from a "simple single-layer perceptron" to "complex multilayer networks" [1].

9.2.3.1 Single-layer neural network: The perceptron

The most fundamental architecture of a neural network is called the perceptron. The perceptron includes two layers of nodes; the input nodes and a single output node. Here, the number of input nodes is equal to the dimensionality d of the underlying data. Each of these input nodes takes and transmit a single numerical attribute to the output node. For that reason, the input nodes just transmit input values and do not apply any computing on these values. In the main perceptron model. The output node is the only node that applies a mathematical function on its inputs. The individual properties in the training data are supposed to be numerical. For simplicity of further discussion, it will be supposed that all input variables are numerical [1].

The function obtained by the perceptron is called *the activation function*, which is a signed linear function [1]. Let $\bar{W} = (w_1, ..., w_d)$ be the weights for the links of d different inputs to the output neuron for a data record of dimensionality d. Furthermore, a bias b is harnessed to the activation function [1]. The output $z_i \in \{-1, +1\}$ for the feature set $(x_1{}^d, ..., x_i{}^d)$ of the ith data record $\bar{X}_i$, is as follows [1]:

$$z_i = \text{sign}\{\sum_{j=1}^{d} w_j x_i{}^j + b\}, \tag{9.13}$$

$$= \text{sign}\{\bar{W}.\bar{X}_i + b\}. \tag{9.14}$$

Here, z_i is the estimation of the perceptron for the class variable of $\bar{X}_i$. Thus, it is preferred to learn the weights so that the value of z_i is equal to y_i for as many training instances as possible. The error in estimation $(z_i y_i)$ could be receive on any of the values of 2, 0, or +2. When the estimated class is true, a value of zero is reached. The main purpose in the neural network algorithms is to learn the vector of weights $\bar{W}$ and bias b. In this way the difference between z_i and the true class variable y_i becomes as smallest as possible [1].

The fundamental perceptron algorithm begins with a random vector of weights. Afterwards, the algorithm feeds the input data items $\bar{X}_i$ into the neural network one by one to produce the estimation z_i. Then, based on the error value $(z_i - y_i)$, the weights are renewed. In particular, the weight vector W_t is renewed, when the data point $\bar{X}_i$ is fed into it in the tth iteration [1]. It is seen as below [1]:

$$\bar{W}^{t+1} = \bar{W}^t + \eta(y_i - z_i)\bar{X}_i. \tag{9.15}$$

Here, η represents the learning rate of the neural network. The perceptron algorithm recursively cycles among all the training examples in the data and recurrently arranges the weights until convergence is attained [1]. A single training data point can be cycled time after time. Here, each cycle is called an *epoch* [1].

Smaller values of η end in a convergence to higher-quality solutions. However, the convergence is slow [1]. In operation, in beginning, the value of η is selected to be large and step by step decreased as the weights approximate to their optimal values [1].

9.2.3.2 Multilayer neural networks

In addition to the input and output layers, multilayer neural networks contain a hidden layer. The nodes in the hidden layer can, in principle, be linked with variable types of topologies [1].

To illustrate, the hidden layer could itself be composed of multiple layers. Additionally, nodes in one layer can feed into nodes of the next layer. In fact, these nodes are assumed to be linked to the nodes in the next layer. This is called the *multilayer feed-forward network* [1]. Because of this assumption, after the number of layers and the number of nodes in each layer are determined by the analyst the topology of the multilayer feed-forward network is automatically specified [1]. The basic perceptron can be seen as a single-layer feed-forward network. A well-known preferred model is one in which a multilayer feed-forward network includes just a single hidden layer. A network such as this could be thought a two-layer feed-forward network [1]. In addition to the properties of the multilayer feed-forward network is that it is not limited to the application of linear signed functions of the inputs. Arbitrary functions named as the logistic, sigmoid, or hyperbolic tangents can be applied in variable nodes of the hidden layer and output layer [1].

To give an example to this kind of function, the training tuple $\bar{X}_i = (x_i{}^1, ..., x_i{}^d)$, yield, an output value of z_i, as below [1]:

$$z_i = \sum_{j=1}^{d} w_j \frac{1}{1+e^{x_i{}^j}} + b. \tag{9.16}$$

If there is a function which is computed at the hidden layer nodes, then the value of z_i can not be an estimated output of the final class label in $\{1, +1\}$ any more. Afterwards, this output is expansed forward to the next layer [1].

Since the expected output of the output node is known to be equal to the training label value, the training process is relatively straightforward in the single-layer neural network. This known as "ground truth", it is used to build an optimization problem in the least squares form, and to renew the weights with a "gradient-descent method". Since the output node is the only neuron with weights in a single-layer network, the renewed process is simple to apply. For the multilayer networks, since there are no training labels related to the outputs of these nodes, the ground-truth output of the hidden layer nodes are not known. For that reason, the weights of these nodes should be calculated when a training example is applied inaccurately. Obviously, when an error is obtained, different kind of "feedback" is needed from the nodes in the forward layers to the nodes in earlier layers about the expected outputs and related errors. This process is managed with the application of the backpropagation algorithm [1]. The back-propagation algorithm includes two main phases [1]:

Forward phase: Here, the inputs for a training process are fed into the neural network. This results in a forward step of computations across the layers, exercising the current set of weights. The ultimate estimated output might be contrasted to the class label of the training process, to control whether the estimated label is an error or not [1].

Backward phase: The basic purpose of this phase is to find out weights in the backward direction by supplying an error estimate of the output of a node in the former layers from the errors in latter layers. In the hidden layer, the error estimate of a node is calculated as a function of the error estimates and weights of the nodes in front of the layer. This is then used to compute an error gradient with respect to the weights in the node and to update the weights of this node. The original update equation is similar to the basic perceptron at a cognitive level. There are a couple of differences that occur because of the nonlinear functions, that are generally applied in hidden layer nodes and errors at these nodes are mostly estimated via "backpropagation", instead of directly estimated by the comparison of the output to a training label [1]. This whole process is outspread backward to renew the weights of all the nodes in the network [1].

An example of a perceptron and a two-layer feed-forward network is illustrated as follows [1]:

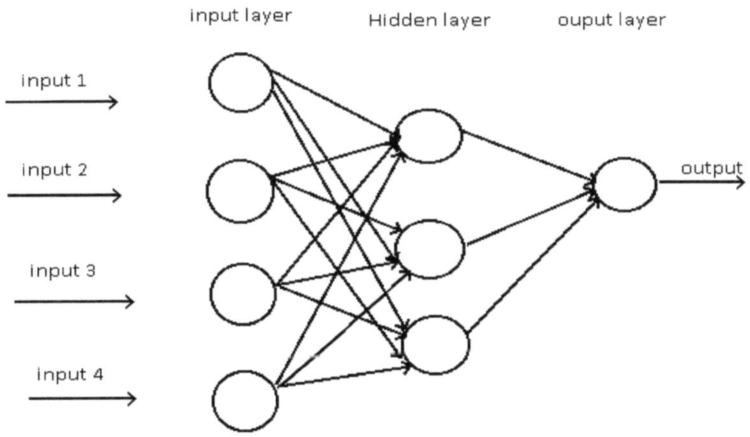

Figure 9.2: Single and multilayer neural networks.

9.3 Application of Machine Learning Techniques into Investor Sentiment

9.3.1 Human factor: investor sentiment

Many financial decisions highly affect people's lives, and these decisions are made at various stages in the economy [12]. Investors' financial performances tend to grieve when they are emotionally reactive which leads to poor impulse control. With the help of some techniques, it can provide a determination of the regions in the brain that are quite active in the course of decision-making processes and, especially, decision making in finance [16, 28, 30]. Economists consider several empirical facts regarding the behavior of individual investors, such as how their emotions and opinions influence their decisions; this constitutes the subject of Behavioral Finance [8, 13, 16]. Feelings can influence thinking styles. A positive mood brings about more creative solutions but they might be risky, whereas, a negative mood may lead to more prudent thinking. These kinds of optimistic or pessimistic expectations can persist and affect asset prices for significant periods, eventually leading to crises [6, 33]. The general name of all these emotional situation is called *Sentiment* as presented previously. The Sentiment term is considered in various ways by financial analysts, academic researchers, and the media, many of them come to terms that Investor Sentiment is economically important. Some researchers address Investor Sentiment as a tendency to trade on noise instead of information. In other respects, some of them consider the Sentiment term as "investor optimism or pessimism". This term is related to feelings; therefore, the media sometimes addresses it as *investor fear* or *risk-aversion* [15, 32].

While investigating assumptions and descriptions about Sentiment, it is seen that it may be possible to measure the Sentiment. There are several Sentiment measures, which vary from measures derived for an academic article to daily indices, used for opposite aims by options traders [15].

There exist some widely used Sentiment measures: "Closed-end fund discount consumer confidence indices, Investor intelligence surveys, Market liquidity, Implied volatility of index options, Ratio of odd-lot sales to purchases, and Net mutual fund redemptions". These measures demonstrate the fact that there is no clear, general explanation of Sentiment by both academic researchers and professional traders. The former employs Investor Sentiment measures to make discussions in favor of market efficiency or to tell the reason and the result of the specific movements. The latter adopts Sentiment indicators as a probable trading tool [15, 32]. Thus, there are conflicts about how to measure it. Two approaches are examined for measuring sentiment "indirect market-based proxies for sentiment" and "direct survey data". The first one is the *market-based* approach, which aims to accumulate Sentiment from financial proxies indirectly. To illustrate this, the put-call ratio and closed-end fund discount are widely mentioned as proxies for sentiment obtained from market reactions. The second approach is to *measure sentiment directly*, employing and benefitting from surveys and questionnaires given to investors. Yale School of Management's Stock Market Confidence Index (YSMI), University of Michigan's Consumer Confidence Index (MCCI), CFA Institute's Global Market Sentiment Survey, and American Association of Individual Investors' (AAII) weekly poll of Investor Sentiment can be given as examples [15, 32]. In order to measure the Sentiment, instead of using a single measure of Investment Sentiment, Baker and Wurgler (2006) [4] prefer to construct an index by using 10 proxies. Inputs that are used for Sentiment Index can be listed as follows: the closed-end fund discount, NYSE share turnover, the number and average first-day returns on initial public offerings (ipo)s, the equity share in new issues (s), and the dividend premium, industrial production index (indpro), nominal durables consumption (consdur), nominal nondurables consumption (consnon), nominal services consumption (consserve), NBER recession indicator (recess), employment (employ), and consumer price index (cpi) [4, 5, 15].

The economic and financial wellness of countries is highly based on the behavior and sentiment of those countrys' financial sector. Well-functioning financial institutions are a fundamental construction of economic development. The influence of these financial institutions and their role in improving economics play a significant role in building the finance industry in countries [3]. The fields of financial mathematics, stochastic calculus, numerical analyses, and scientific computing are combined to solve the actual issues in the financial industry too. The application of these computational techniques to finance and the sentiment of investors is important for the business world to create and investigate strategic planning by providing perspective into what be the future if a strategy is performed, and forecasting the risks related to financial instruments [14]. Even though on a large scale, the use of traditional statistical approaches in constructing predictions in financial and investor sentiment fields occur these approaches can enhance some restrictive assumptions such as linearity, normality, independence between independent, and dependent variables. Therefore, because of these fields' stochastic and complex nature, it is a compelling task to forecast with the help of classical statistical techniques [2, 20, 22]. That is why, these traditional models have given rise to a growing interest in machine learning techniques [3]. Numerous researchers examine a lot of techniques from machine learning and data mining to solve problems in finance [14]. Machine learning techniques success is based on their property to model nonlinear systems with minimal initial assumptions and advanced forecasting accuracy [25]. Many investors' decisions and opinions in the fields of the financial industry are researched and applied by machine learning techniques. This provides help in the finance industry to comprehend the data and obtain a competitive advantage from the data [14]. On the other hand, hybrid models which combine several machine learning techniques tend to improve more and more [14, 9]. The association derived by machine learning has proved prosperous for hybrid methods. However, the process should always take into consideration the significance of data quality, as the ambiguity of data always requires the robustness of the machine learning technology [14].

In this regard, different methods are used in the literature via either a single machine learning model or hybrid machine learning model: Bahrammirzaee (2010) studies a comparative research review about three well-known artificial intelligence techniques in the financial market; which are artificial neural networks (ANN), expert systems, and hybrid intelligence systems. In his study, the financial market is classified and explained in three aspects, namely, credit evaluation, portfolio management, financial prediction, and planning. This study shows that artificial intelligent techniques obtain more accurate results than the traditional statistical methods in dealing with financial problems [3]. Lu et al. (2010) [22] compare the forecasting performance of MARS, backpropagation neural network (BPN), support vector regression (SVR), and multiple linear regression (MLR) models in the Shanghai B-Share stock index in order to forecast stock index prices. Experimental results indicate that MARS outperforms BPN, SVR, and MLR in terms of prediction error and prediction accuracy [22]. Furthermore, Lin (2010) [21] proposes a new approach based on the two-stage hybrid model of the logistic regression-ANN for the construction of a financial distress warning system in the banking industry while emerging market between 1998–2006. This study considers a two-stage hybrid model integrated to the benefits of logistic regression and ANN while preventing computational complexity. Some innovative treatments are adopted so that the proposed approach can be found to outperform traditional models. The results indicate that the factors of liquidity, capital, and asset quality are significantly related to the financial distress of banks in the emerging market. In the prediction of financially distressed banks, the proposed two-stage hybrid model provides the best fit based on the Root Mean Square Error ($RMSE$) and R-Squared (R^2) measures and demonstrates an enhanced prediction power than the conventional ones [21]. Kao et al. (2013) [17] suggest a new stock price forecasting model which includes wavelet transform, multivariate adaptive regression splines (MARS), and support vector regression (SVR) (called Wavelet-MARS-SVR) to improve the forecast accuracy. The performance of these underlying methods is utilized by comparing the forecasting results of Wavelet-MARS-SVR with the ones made by other five competing approaches, namely, (Wavelet-SVR, Wavelet-MARS, single ARIMA, single SVR, and single ANFIS). The results of this study show that the proposed approach is superior to other competing models [17]. Jadhav et al. (2017) [14] perform research on data mining techniques for the financial industry from 2010 to 2015. According to reviews, stock prediction and Credit rating receive the most interest from researchers compared to loan prediction, money laundering, and time series prediction. On the other hand, the nonlinear mapping techniques are studied more than linear techniques because of the dynamics, uncertainty and variety of data. Additionally, in this study, it is shown that hybrid methods have more accurate prediction, closely followed by the neural network technique [14].

9.3.2 Application

After all these studies mentioned above, in this study indicates the applicability of three machine learning techniques by themselves and also by the hybrid model to forecast the sentiment level of investors. The purpose of this study is to compare the forecasting performances of MARS, neural network (NN), random forest (RF) models with each other and then by constructing models such as MARS-NN, MARS-RF, RF-MARS, RF-NN, NN-MARS, and MARS-NN to predict sentiment levels of investors. To evaluate the performance of these approaches, the Sentiment index which is constructed by Baker and Wurgler (2006) [4] using 6 proxies is adopted. While constructing this index, raw data are produced by 10 proxies. Then, these raw data are used for the application of machine learning techniques. Therefore, inputs that are implemented for the Sentiment Index are listed as the closed-end fund discount, NYSE share turnover, the number, and average first-day returns on IPO (initial public offerings)s, the equity which shares new issues, the dividend premium, indpro (industrial production index), consdur (nominal durables consumption), consnon (nominal nondurables consumption), consserv (nominal services consumption), recess (NBER recession indicator), employ (employment), and cpi (consumer price index) [4, 5, 15]. Finally, these data are monthly and in this study, it is used from the year 2000 to the end of 2018.

To evaluate the accuracy of the model, we need to know the meanings of some statistical measures:

(i) *Mean Squared Error (MSE)*: the average of the squared difference between the real and predicted values in the data set.

(ii) *R-Squared (RSq)*: a measure of how well the model fits the training data [10].

(iii) *Generalized R-Squared (RSq)*: an estimate of the predictive power of the model (calculated over all responses) [10].

Hereby, at first, we begin our analyses with forecasting by the MARS method. There are 10 inputs which are mentioned above and 1 output which is sentiment itself. For this data set, using the *R* Programme, the highest degree of interactions is taken as 3 since this is the most preferable interaction level and the number of basis functions is set to 25. As a result, we obtain generalized cross validation (GCV) values as 0.015, GRSq as 0.968, RSq as 0.978, and finally, the mean square error as 0.221. To apply a two-stage case, we ignore less important variables. According to importance of variables. we eliminate four variables, namely, nipo, s, consdur, consnon. Then, we take the remaining data set for the RF model and we get RSq as 0.954 and MSE as 0.022 (MARS-RF). We also perform this remaining data set into NN model (MARS-NN). Here, the MSE is obtained as 0.41.

Secondly, we apply the RF model for the prediction of Sentiment. We use the same data. The results are computed as 0.948 for the RSq and 0.025 for MSE. Afterwards, to apply a two-stage case, we ignore less important variables. After discarding these variables which are s and nipo, initially, we merely take the remaining data set for the MARS model (RF-MARS). The results are obtained as 0.016 for GCV, 0.968 for GRSq, 0.978 for RSq, and 0.265 for MSE. On the other hand, after discarding less important variables according to the RF model, this time we apply the remaining data for the NN model. The MSE is computed as 0.0488 (RF-NN).

Thirdly, we apply the NN model to forecast Sentiment. The application is done with 90% for the training and 10% for the test period. The MSE is found as 0.034. According to the estimated weights of variables, only employ and conssserve variables remain (they are the only ones with value 1). Firstly, we implement the remaining data for the RF model and the results are calculated as 0.948 for RSq and 0.025 for MSE. As a second case, we apply the remaining data set from NN model into the MARS model. The results are obtained as 0.021 for GCV, 0.956 for GRSq, 0.969 for RSq, and finally, 1.129 for MSE (NN-MARS). These outcomes under distinct models are also presented in Tables 9.1, 9.2 and 9.3.

According to Table 9.1, it is seen that we have better outcomes by using only the MARS method. Moreover, its GCV is less than the hybrid models and it has higher GRSq and RSq values than the hybrid models.

As demonstrated in Table 9.2, the model MARS-RF has better results compared to other models. This proposed model gives lower MSE and higher RSq values.

Table 9.1: The result of MARS and methods whose second stage is MARS.

Methods	RSS	GCV	GRSq	RSq	MSE
MARS	2.427	0.015	0.968	0.978	0.22
RF-MARS	2.387	0.016	0.968	0.9781	0.265
NN-MARS	3.388	0.021	0.956	0.969	1.129

Table 9.3 shows that by using only the NN method, we achieve much better results.

To conclude, we can say that the MARS model itself outperforms better compared to the other hybrid models. On the other hand, the hybrid model, which is MARS-RF, also has better results with respect to the single RF model and the NN-RF model. Using MARS as a first-stage modeling tool with the obtained results being the inputs to RF contributes to the achievement of the model [18].

Table 9.2: The result of RF and methods whose second stage is RF.

Methods	MSE	RSq
RF	0.025	0.948
MARS-RF	0.022	0.954
NN-RF	0.025	0.948

Table 9.3: The result of NN and methods whose second stage is NN.

Methods	MSE
NN	0.034
MARS-NN	0.416
RF-NN	0.438

The single NN model managed less error values than hybrid models. As stated before, the process should always take into consideration the importance of data structure. According to the general results, the MARS single model, NN single model and MARS-RF hybrid model achieve better results in this kind of sentiment data. Therefore, their model selection criteria work better regarding other models.

9.4 Conclusion

Various recent real financial applications have nonlinear and complex behaviors. Since classical statistical methods depend on some restrictive assumptions and applications, methods are required to deal with these stochastic problems [2, 3].

Machine learning methods are well-known and beneficial tools for prediction problems and have already been successfully applied to numerous financial associated problems. In this study, apart from pure financial related problems, we focus on the behavior of financial problems which is based on the investors' behavior introduced as Sentiment.

Hence, the goal of this study is to compare the forecasting performance of the sentiment index by using single MARS, RF, NN models, and MARS-NN, MARS-RF, RF-MARS, RF-NN, NN-MARS, and NN-RF hybrid models.

Therefore, the results show that MARS single model, NN single model, and MARS-RF hybrid model outperform better with this kind of sentiment data.

For future studies, we might work with another behavioral data that shows the investors' sentiment levels. We could also add many other machine learning techniques such as support vector machine, hidden markov model, CMARS, etc.

Operations research and machine learning can make use of each other. One of the important advantages is that the relation between them raises the applications of operational research, especially, in data science through the employment of machine learning [29].

References

[1] C.C. Aggarwal. *Data Mining: The Textbook*. Springer, 2015.

[2] B.S. Arasu, M. Jeevananthan, N. Thamaraiselvan and B. Janarthanan. Performances of data mining techniques in forecasting stock index–evidence from India and us. *Journal of the National Science Foundation of Sri Lanka*, 42(2): 177–191, 2014.

[3] A. Bahrammirzaee. A comparative survey of artificial intelligence applications in finance: Artificial neural networks, expert system and hybrid intelligent systems. *Neural Computing and Applications*, 19(8): 1165–1195, 2010.

[4] M. Baker and J. Wurgler. Investor sentiment and the cross-section of stock returns. *The Journal of Finance*, 61(4): 1645–1680, 2006.

[5] M. Baker and J. Wurgler. Investor sentiment in the stock market. *The Journal of Economic Perspectives*, 21(2): 129–151, 2007.

[6] S.-K. Bormann. Sentiment indices on financial markets: What do they measure? Technical report, Economics Discussion Papers, 2013.

[7] L. Breiman, J. Friedman, R. Olshen and C. Stone. *Classification and Regression Trees*. Wadsworth and Brooks, Monterey, CA, 1984.

[8] D. Colander. *The Complexity Vision and the Teaching of Economics*. Edward Elgar Publishing, 2000.

[9] J. De Andrés, P. Lorca, F.J. de Cos Juez and F. Sánchez-Lasheras. Bankruptcy forecasting: A hybrid approach using fuzzy c-means clustering and multivariate adaptive regression splines (mars). *Expert Systems with Applications*, 38(3): 1866–1875, 2011.

[10] N.I. Dorf and R. Documentation. Find the optimal testing configuration for non-informative two-stage hierarchical testing.

[11] J. Friedman, T. Hastie and R. Tibshirani. *The Elements of Statistical Learning*, volume 1. Springer series in statistics Springer, Berlin, 2001.

[12] C. Frydman and C.F. Camerer. The psychology and neuroscience of financial decision making. *Trends in Cognitive Sciences*, 20(9): 661–675, 2016.

[13] C.D. Frydman. *Essays in Neurofinance*. PhD thesis, California Institute of Technology, 2012.

[14] S. Jadhav, H. He and K.W. Jenkins. An academic review: Applications of data mining techniques in finance industry. 2017.

[15] B. Kalaycı. *Identification of Coupled Systems of Stochastic Differential Equations in Finance including Investor Sentiment by Multivariate Adaptive Regression Splines*. PhD thesis, Middle East Technical University, 2017.

[16] B. Kalaycı, A. Özmen and G.-W. Weber. Mutual relevance of investor sentiment and finance by modeling coupled stochastic systems with mars. *Annals of Operations Research*, pp. 1–24, 2020.

[17] L.-J. Kao, C.-C. Chiu, C.-J. Lu and C.-H. Chang. A hybrid approach by integrating wavelet-based feature extraction with mars and svr for stock index forecasting. *Decision Support Systems*, 54(3): 1228–1244, 2013.

[18] T.-S. Lee and I.-F. Chen. A two-stage hybrid credit scoring model using artificial neural networks and multivariate adaptive regression splines. *Expert Systems with Applications*, 28(4): 743–752, 2005.

[19] T.-S. Lee, C.-C. Chiu, Y.-C. Chou and C.-J. Lu. Mining the customer credit using classification and regression tree and multivariate adaptive regression splines. *Computational Statistics & Data Analysis*, 50(4): 1113–1130, 2006.

[20] T.-S. Lee and C.C. Yang. Incorporating financial ratios and intellectual capital in bankruptcy predictions. In *Proceedings of the National Taiwan University International Conference in Finance, Taiwan, December*, pp. 20–21. Citeseer, 2004.

[21] S.L. Lin. A two-stage logistic regression-ann model for the prediction of distress banks: Evidence from 11 emerging countries. *African Journal of Business Management*, 4(14): 3149–3168, 2010.

[22] C.-J. Lu, C.-H. Chang, C.-Y. Chen, C.-C. Chiu and T.-S. Lee. Stock index prediction: A comparison of mars, bpn and svr in an emerging market. In *2009 IEEE International conference on Industrial Engineering and Engineering Management*, pp. 2343–2347. IEEE, 2009.

[23] A. Özmen. *Robust Optimization of Spline Models and Complex Regulatory Networks*. Springer, 2016.

[24] A. Özmen. *Robust Conic Quadratic Programming in Applied to Quality Improvement—A Robustification of CMARS*. MSC thesis at Middle East Technical University, Ankara, Turkey, September, 2010.

[25] V. Plakandaras, T. Papadimitriou and P. Gogas. Forecasting daily and monthly exchange rates with machine learning techniques. *Journal of Forecasting*, 34(7): 560–573, 2015.

[26] D. Seçilmiş. *Deterministic Modeling and Inference of Biochemical Networks*. PhD thesis, Middle East Technical University, 2017.

[27] P. Sephton. Forecasting recessions: Can we do better on MARS. *Review*, 83, 2001.

[28] M.Z. Shariff, J. Al-Khasawneh and M. Al-Mutawa. Risk and reward: A neurofinance perspective. *International Review of Business Research Papers*, 8(6): 126–133, 2012.

[29] A. Talavera and A. Luna. Operational research and machine learning: An engineering course. In *2019 IEEE World Conference on Engineering Education (EDUNINE)*, pp. 1–5. IEEE, 2019.

[30] K.C. Tseng. Behavioral finance, bounded rationality, neuro-finance, and traditional finance. *Investment Management and Financial Innovations*, (3, Iss. 4): 7–18, 2006.

[31] D. Yao, J. Yang and X. Zhan. A novel method for disease prediction: Hybrid of random forest and multivariate adaptive regression splines. *Journal of Computers*, 8(1):170–177, 2013.

[32] C. Zhang. Defining, modeling, and measuring investor sentiment. *University of California, Berkeley, Department of Economics*, 2008.

[33] M. Zouaoui, G. Nouyrigat and F. Beer. How does investor sentiment affect stock market crises evidence from panel data. *Financial Review*, 46(4): 723–747, 2011.

IV

OPERATION RESEARCH IN MEDICAL APPLICATION

Chapter 10

An Algorithm and Stability Approach for the Acute Inflammatory Response Dynamic Model

Burcu Gürbüz[1,]* and *Aytül Gökçe*[2]

10.1 Introduction

Acute inflammatory response that starts shortly after a specific trauma is the initial reaction of the body to get rid of harmful stimuli and pathogens [5, 9, 31]. In general, the inflammatory response wipes out the pathogens from the body and repairs the unhealthy cells. However, if the inflammation is systemic and uncontrolled due to infection, the body's response may be also be damaging to healthy cells, leading to an inappropriate inflammatory response [24, 27]. This excessive response of the whole body to trauma is clinically called sepsis, which may trigger damage in tissue, dysfunction in organs and even death [6, 27]. In recent years, much information about molecular structure and signalling pathways of innate inflammatory response has been obtained, though available therapies are not efficient for a definitive treatment. Therefore, mathematical models may provide essential insights into the dynamics of inflammatory response and

[1] Johannes Gutenberg-University Mainz.
[2] Ordu University.
* Corresponding author: burcu.gurbuz@uni-mainz.de

contribute to the understanding of the underlying interactions between various biological components [24].

Several mathematical models have been developed and a rich repertoire of interesting results have been obtained to investigate the basic picture of inflammatory response. A simple non-linear mathematical model that includes three variables, i.e., pathogen, an early-inflammatory mediator and a late-inflammatory mediator, have been proposed by Kumar et al. (2006) [24], who analysed various clinically relevant scenarios with different model parameters and discussed that sepsis as a complicated disease should be targeted with multi-directional interventions. Later, Reynolds et al. [31] developed a more advanced model to explore the importance of dynamic anti-inflammation on the pathogenic infection, depending on variables pathogen, activated phagocytes, tissue damage, and anti-inflammatory mediator. Substituting the pathogen variable with an endotoxin variable in [31], Day et al. (2006) [9] analysed a mathematical model incorporating the role of an endotoxin molecule inducing acute inflammatory response in the immune system.

The term reliability usually refers to the probability that a component or system will operate satisfactorily either at any particular instant at which it is required or for a certain length of time. Fundamental to quantifying reliability is a knowledge of how to define, assess and combine probabilities [4]. This may hinge on identifying the form of the variability which is inherent in most processes. If all components had a fixed known lifetime there would be no need to model reliability.

Nonlinear dynamics is of great importance in many different branches of engineering and applied sciences. In the most common scenarios, the nonlinearity of the dynamical structures express the real-world phenomena and describe the complicated conceptions in details. Thus, the mathematical approach leading the complex modeling themes includes some parameters, derivatives, and delay terms. Then, we consider the numerical methodology and qualitative analyses in detail to investigate these systems of nonlinear differential equations comprehensively. However, numerical approximations to these models are not straightforward most of the time. Thus, different numerical methods are improved and modified to obtain an approximation to these problems. Particularly, there exist some research studies as follows. For example, the Adomian decomposition method and Runge-Kutta method have been considered by Shawagfeh and Kaya [34], series of exponential functions have introduced by Wasow [37], Adaptive Runge-Kutta Method has been described by Chowdhury et al. (2019) [7], Sinc-Derivative interpolation have been explained by Abdella and Trivedi (2020) [1], the matrix methods have been implemented by Yüzbaşı and Sezer (2016) (cf. [39, 40]) and Laguerre collocation has been introduced by Gürbüz et al. (see [16, 17, 19, 21]).

Accordingly, the organization of our study is as follows. In Section 10.2, the mathematical model is introduced with details. In Section 10.3, the numerical method is introduced clearly where its accuracy is also addressed comprehensively by the convergence and error bounds theorems. The stability analysis of the model is presented in Section 10.4. In Section 10.5, numerical simulations are given together with the accuracy results. Conclusion remarks, outlook for the further studies are given and also discussion on the results are introduced in Section 10.6.

10.2 Mathematical Model

In this section, we introduce the mathematical model which includes three variables: u_1 is a pathogen as an instigator of the innate immune response, u_2 is an early pro-inflammatory mediator and u_3 is a late pro-inflammatory feedback [24]. The model presenting the acute inflammatory response with the dynamics is

$$\frac{du_1}{dt} = \phi u_1(1-u_1) - \mu u_1 u_2, \tag{10.1}$$

$$\frac{du_2}{dt} = (\eta u_1 + u_3)u_2(1-u_2) - u_2, \tag{10.2}$$

$$\frac{du_3}{dt} = \zeta f(u_2) - \xi u_3, \tag{10.3}$$

where

$$f(u_2) = 1 + \tanh\left(\frac{u_2 - \theta}{\psi}\right). \tag{10.4}$$

In this final expression, an activation threshold and an activation width are given as θ and ψ, respectively. Here, all the parameters and the unknowns of the dynamic are nonnegative.

While parameters are fixed to $\phi = 3, \eta = 25, \zeta = 15, \xi = 1, \theta = 1, \psi = 0.5$,

- For unstable dynamics (see Figure 1 in the ref paper) set: $\mu = 30$ with initial conditions, such $u_1(0) = 0.01$, $u_2(0) = 0.05$, $u_3(0) = 0.539$ elsewhere.
- For stable dynamics (see Figure 3 in the ref paper) set : $\mu = 3$ with initial conditions, such $u_1(0) = 0.01$, $u_2(0) = 0.05$, $u_3(0) = 0.539$ elsewhere.

10.3 Numerical Method

In this section, we introduce a numerical method that investigates the numerical solutions of the model defined in 10.1–10.4 together with the initial conditions. The method transforms the problem into the matrix equations and the collocation points are applied pointwise for finding the solutions at the given interval. As a first step, we consider the fundamental matrix relations which is to obtain a general matrix representation of the model. Then, we find an augmented matrix form for our problem with the help of the collocation based approach. Thus, it gives us an understanding of the polynomial solution of the problem as an alternative.

As a part of this study, the numerical solution of the present model is obtained by the Taylor series. Firstly, we organise the system of differential equations to obtain the matrix forms of the equations. Then, we describe the numerical technique for the collocation points. Secondly, the initial conditions are also shown by their matrix representations. Here, the approximate solution of the system is described in the Taylor series as follows:

$$u_i(t) \cong u_{iN}(t) = \sum_{n=0}^{N} u_{in}(t)(t-c)^n, \tag{10.5}$$

where

$$u_{in}(t) = \frac{u_{in}^{(n)}(c)}{n!}, \quad i = 1,2,3, \quad a \leq t \leq b. \tag{10.6}$$

Moreover, we organise the Equation 10.4 by using the N-th order of Taylor series form as below.

$$\begin{aligned} f(u_2) &= 1+\tanh\left(\frac{u_2-\theta}{\psi}\right) = \left(\tanh\left(\frac{f(c)-\theta}{\psi}\right)+1\right) + \frac{u_2\left(f'(c)-f'(c)\tanh^2\left(\frac{f(x)\theta}{\psi}\right)\right)}{\psi} \\ &+\frac{1}{2\psi^2}u_2^2\left(-\psi f''(c)\tanh^2\left(\frac{f(c)-\theta}{\psi}\right)+2f'(c)^2\tanh^3\left(\frac{f(c)-\theta}{\psi}\right)-2f'(c)^2\right. \\ &\left.\tanh\left(\frac{f(c)-\theta}{\psi}\right)+\psi f''(c)\right)+\frac{1}{6\theta^3}u_2^3\left(-\psi^2 f^{(3)}(c)\tanh^2\left(\frac{f(c)-\theta}{\psi}\right)\right. \\ &-6f'(c)^3\tanh^4\left(\frac{f(c)-\theta}{\psi}\right)+6\psi f'(c)f''(c)+8f'(c)^3\tanh^2\left(\frac{f(c)-\theta}{\psi}\right) \\ &\left.\tanh^3\left(\frac{f(c)-\theta}{\psi}\right)-6\psi f'(c)f''(c)\tanh\left(\frac{f(c)-\theta}{\psi}\right)+\psi^2 f^{(3)}(c)-2f'(c)^3\right) \\ &+\frac{1}{24\theta^4}u_2^4\left(-\psi^3 f^{(4)}(c)\tanh^2\left(\frac{f(c)-\theta}{\psi}\right)+6\psi^2 f''(c)^2\tanh^3\left(\frac{f(c)-\theta}{\psi}\right)\right. \\ &-6\psi^2 f''(c)^2\tanh\left(\frac{f(c)-\theta}{\psi}\right)+8\psi^2 f^{(3)}(c)f'(c)\tanh^3\left(\frac{f(c)-\theta}{\psi}\right) \\ &-8\psi^2 f^{(3)}(c)f'(c)\tanh\left(\frac{f(c)-\theta}{\psi}\right)+24f'(c)^{24}\tanh^5\left(\frac{f(c)-\theta}{\psi}\right) \\ &-40f'(c)^4\tanh^3\left(\frac{f(c)-\theta}{\psi}\right)+16f'(c)^4\tanh\left(\frac{f(c)-\theta}{\psi}\right) \\ &-36\psi f'(c)^2 f''(c)^2\tanh^4\left(\frac{f(c)-\theta}{\psi}\right)+48\psi f'(c)^2 f''(c)\tanh^2\left(\frac{f(c)-\theta}{\psi}\right) \\ &\left.+\psi^3 f^{(4)}(c)-16\psi f'(c)^2 f''(c)\right)+\cdots+O(u_2^N) = f(u_{2,N}). \end{aligned} \tag{10.7}$$

10.3.1 Fundamental matrix relations

Now we consider the general and implicit form of the model for $c = 0$. Then, we represent the matrix form as follows:

$$[u_i(t)] = \mathbf{T}(t)\mathbf{U}_i, \quad i = 1,2,3, \tag{10.8}$$

where we have the following matrices

$$\begin{aligned} \mathbf{T}(t) &= \begin{bmatrix} 1 & t & t^2 & \cdots & t^N \end{bmatrix}, \\ \mathbf{U}_i &= \begin{bmatrix} u_{i0} & u_{i1} & u_{i2} & \cdots & u_{iN} \end{bmatrix}^T. \end{aligned}$$

The system of differential equations includes the first derivatives of the unknown functions, $u_1(t)$, $u_2(t)$ and $u_3(t)$, as $u_1'(t)$, $u_2'(t)$ and $u_3'(t)$, respectively ([16, 19]). Their matrix representations are shown by

$$[u_i'(t)] = \mathbf{T}(t)\mathbf{B}\mathbf{U}_i, \quad i = 1,2,3, \tag{10.9}$$

where

$$\mathbf{B} = \begin{bmatrix} 0 & 1 & 0 & \cdots & 0 \\ 0 & 0 & 2 & \cdots & 0 \\ \vdots & \vdots & \vdots & \ddots & \vdots \\ 0 & 0 & 0 & \cdots & N \\ 0 & 0 & 0 & \cdots & 0 \end{bmatrix}.$$

Now nonlinear terms of the system 10.1–10.4 are shown by the matrices. Nonlinear terms are presented with regard to the explicit form of the system. Then, initially, we have;

$$[u_1^2(t)] = \mathbf{T}(t)\overline{\mathbf{T}}(t)\overline{\mathbf{U}}_{11}, \tag{10.10}$$

where

$$\begin{aligned} \overline{\mathbf{T}}(t) &= \begin{bmatrix} \mathbf{T}(t) & 0 & \cdots & 0 \\ 0 & \mathbf{T}(t) & \cdots & 0 \\ \vdots & \vdots & \ddots & \vdots \\ 0 & 0 & \cdots & \mathbf{T}(t) \end{bmatrix}, \\ \overline{\mathbf{U}}_{11} &= \begin{bmatrix} u_{10}\mathbf{U}_1 & u_{11}\mathbf{U}_1 & \cdots & u_{1N}\mathbf{U}_1 \end{bmatrix}^T. \end{aligned}$$

We also describe the matrix relation of the following nonlinear terms in the explicit form of the system 10.1–10.4 as

$$[u_1(t)u_2(t)] = \mathbf{T}(t)\overline{\mathbf{T}}(t)\overline{\mathbf{U}}_{21}, \tag{10.11}$$

where

$$\overline{\mathbf{U}}_{21} = \begin{bmatrix} u_{10}\mathbf{U}_2 & u_{11}\mathbf{U}_2 & \cdots & u_{1N}\mathbf{U}_2 \end{bmatrix}^T.$$

Furthermore, we have

$$[u_2^2(t)] = \mathbf{T}(t)\overline{\mathbf{T}}(t)\overline{\mathbf{U}}_{22}, \tag{10.12}$$

in which

$$\overline{\mathbf{U}}_{22} = \begin{bmatrix} u_{20}\mathbf{U}_2 & u_{21}\mathbf{U}_2 & \cdots & u_{2N}\mathbf{U}_2 \end{bmatrix}^T.$$

Also, we have the following nonlinear terms:

$$[u_2^2(t)u_3(t)] = \mathbf{T}(t)\overline{\mathbf{T}}(t)\overline{\mathbf{T}}^*(t)\overline{\mathbf{U}}_{223}^*, \tag{10.13}$$

where

$$\overline{\mathbf{T}}^*(t) = \begin{bmatrix} \overline{\mathbf{T}}(t) & 0 & \cdots & 0 \\ 0 & \overline{\mathbf{T}}(t) & \cdots & 0 \\ \vdots & \vdots & \ddots & \vdots \\ 0 & 0 & \cdots & \overline{\mathbf{T}}(t) \end{bmatrix},$$

$$\overline{\mathbf{U}}_{223}^* = \begin{bmatrix} u_{30}\overline{\mathbf{U}}_{22} & u_{31}\overline{\mathbf{U}}_{22} & \cdots & u_{3N}\overline{\mathbf{U}}_{22} \end{bmatrix}^T.$$

In these calculations, we also consider the matrix form of the nonlinear term of the system in Equation 10.4 via

$$[f(u_{2,N})] = \mathbf{f}_s(t,\theta,\psi)\mathbf{T}(t)\overline{\mathbf{T}}(t)\overline{\mathbf{U}}_{23}, \tag{10.14}$$

where $s = 0,1,2,3,\ldots,N$ and

$$\mathbf{f}_s(t,\theta,\psi) = \begin{bmatrix} f_0(t,\theta,\psi) & 0 & \cdots & 0 \\ 0 & f_1(t,\theta,\psi) & \cdots & 0 \\ \vdots & \vdots & \ddots & \vdots \\ 0 & 0 & \cdots & f_N(t,\theta,\psi) \end{bmatrix},$$

$$\overline{\mathbf{U}}_{23} = \begin{bmatrix} u_{30}\mathbf{U}_2 & u_{31}\mathbf{U}_2 & \cdots & u_{3N}\mathbf{U}_2 \end{bmatrix}^T.$$

Hence, we can organise the system by replacing the matrix representations 10.8–10.14 as in Equation 10.1–10.4. Thereby,

$$(\mathbf{T}(t)\mathbf{B} - \phi\mathbf{T}(t))\mathbf{U}_1 = -\phi\mathbf{T}(t)\overline{\mathbf{T}}(t)\overline{\mathbf{U}}_{11} - \mu\mathbf{T}(t)\overline{\mathbf{T}}(t)\overline{\mathbf{U}}_{21}, \tag{10.15}$$

$$(\mathbf{T}(t)\mathbf{B} + \mathbf{T}(t))\mathbf{U}_2 = \eta\mathbf{T}(t)\overline{\mathbf{T}}(t)\overline{\mathbf{U}}_{21} + \mathbf{T}(t)\overline{\mathbf{T}}(t)\overline{\mathbf{T}}^*(t)\overline{\mathbf{U}}_{223}^*, \tag{10.16}$$

$$(\mathbf{T}(t)\mathbf{B} + \xi\mathbf{T}(t))\mathbf{U}_3 = \zeta\mathbf{f}_s(t,\theta,\psi)\mathbf{T}(t)\overline{\mathbf{T}}(t)\overline{\mathbf{U}}_{23}. \tag{10.17}$$

Hereby, we show the matrix system of equations in the form below.

$$\mathbf{P}_i(t)\mathbf{U}_i + \mathbf{R}_i(t)\overline{\mathbf{U}}_{11} + \mathbf{S}_i(t)\overline{\mathbf{U}}_{21} + \mathbf{M}_i(t)\overline{\mathbf{U}}_{23} + \mathbf{K}_i(t)\overline{\mathbf{U}}_{223}^* = \mathbf{Z}_i(t), \tag{10.18}$$

in which $i = 1,2,3$ or it is shown implicitly as

$$\mathbf{P}_i(t)\mathbf{U}_i - \mathbf{R}_i(t)\overline{\mathbf{U}}_{ij} - \mathbf{S}_i(t)\overline{\mathbf{U}}_{ijk}^* = \mathbf{Z}_i \tag{10.19}$$

where $i, j, k = 1, 2, 3$ and

$$\begin{aligned}
\mathbf{P}_1(t) &= \mathbf{T}(t)\mathbf{B} - \phi\mathbf{T}(t), \quad \mathbf{P}_2(t) = \mathbf{T}(t)\mathbf{B} + \mathbf{T}(t), \quad \mathbf{P}_3(t) = \mathbf{T}(t)\mathbf{B} + \xi\mathbf{T}(t), \\
\mathbf{R}_1(t) &= \phi\mathbf{T}(t)\overline{\mathbf{T}}(t), \quad \mathbf{R}_2(t) = -\eta\mathbf{T}(t)\overline{\mathbf{T}}(t), \quad \mathbf{R}_3(t) = -\zeta\mathbf{f}_s(t,\theta,\psi)\mathbf{T}(t)\overline{\mathbf{T}}(t), \\
\mathbf{S}_1(t) &= \mathbf{0}, \quad \mathbf{S}_2(t) = -\mathbf{T}(t)\overline{\mathbf{T}}(t)\overline{\mathbf{T}}^*(t), \quad \mathbf{S}_3(t) = \mathbf{0}.
\end{aligned}$$

Additionally,

$$\begin{aligned}
\mathbf{P}_i(t) &= \begin{bmatrix} \mathbf{P}_1(t) & 0 & 0 \\ 0 & \mathbf{P}_2(t) & 0 \\ 0 & 0 & \mathbf{P}_3(t) \end{bmatrix}, \\
\mathbf{R}_i(t) &= \begin{bmatrix} \mathbf{R}_1(t) & 0 & 0 \\ 0 & \mathbf{R}_2(t) & 0 \\ 0 & 0 & \mathbf{R}_3(t) \end{bmatrix}, \\
\mathbf{S}_i(t) &= \begin{bmatrix} \mathbf{S}_1(t) & 0 & 0 \\ 0 & \mathbf{S}_2(t) & 0 \\ 0 & 0 & \mathbf{S}_3(t) \end{bmatrix}.
\end{aligned}$$

As a result, we present the fundamental matrix relations that gives us a strong tool for the application of the algorithm. Then they are addressed for the execution of the algorithm.

10.3.2 The collocation approach

In this part, we consider the collocation approach for the implementation of the method. In other words, this is a collocation approach with the multi-step algorithm (see [2, 35]). First we give the definitions of the related arguments for the numerical method as follows (see [17, 21]).

Definition: For every $l \in \mathbb{N}$, we present

$$t_l = a + l\frac{(b-a)}{N}, \qquad l = 1, 2, \ldots, N, \tag{10.20}$$

which are denoted as "*collocation points*".
Now, we apply the collocation points into Equation 10.18

$$\mathbf{P}_i(t_l)\mathbf{U}_i - \mathbf{R}_i(t_l)\overline{\mathbf{U}}_{ij} - \mathbf{S}_i(t_l)\overline{\mathbf{U}}^*_{ijk} = \mathbf{Z}_i, \tag{10.21}$$

where $i, j, k = 1, 2, 3$ and we obtain the system of matrix equations. Moreover, it can be represented shortly by

$$\mathbf{P}_i\mathbf{U}_i - \mathbf{R}_i\overline{\mathbf{U}}_{ij} - \mathbf{S}_i\overline{\mathbf{U}}^*_{ijk} = \mathbf{Z}_i, \quad i, j, k = 1, 2, 3, \tag{10.22}$$

where

$$\mathbf{P}_i(t_l) = \begin{bmatrix} \mathbf{P}_1(t_l) & 0 & 0 \\ 0 & \mathbf{P}_2(t_l) & 0 \\ 0 & 0 & \mathbf{P}_3(t_l) \end{bmatrix}, \mathbf{R}_i(t_l) = \begin{bmatrix} \mathbf{R}_1(t_l) & 0 & 0 \\ 0 & \mathbf{R}_2(t_l) & 0 \\ 0 & 0 & \mathbf{R}_3(t_l) \end{bmatrix},$$

$$\mathbf{S}_i(t_l) = \begin{bmatrix} \mathbf{S}_1(t_l) & 0 & 0 \\ 0 & \mathbf{S}_2(t_l) & 0 \\ 0 & 0 & \mathbf{S}_3(t_l) \end{bmatrix}, \quad i = 1,2,3, \quad l = 1,2,\ldots,N.$$

Similarly, we apply the same procedure to the initial conditions $u_1(0)$, $u_2(0)$ and $u_3(0)$. Their matrix representations are defined by using the Equation 10.8 as below.

$$[u_i(0)] = \mathbf{T}(0)\mathbf{U}_i, \quad i = 1,2,3, \tag{10.23}$$

in which

$$\mathbf{T}(0) = \begin{bmatrix} 1 & 0 & 0 & \cdots & 0 \end{bmatrix},$$
$$\mathbf{U}_i = \begin{bmatrix} u_{i0} & u_{i1} & u_{i2} & \cdots & u_{iN} \end{bmatrix}^T.$$

Here,

$$[u_1(0)] = \mathbf{T}(0)\mathbf{U}_1 = \Gamma_1, \quad [u_2(0)] = \mathbf{T}(0)\mathbf{U}_2 = \Gamma_2, \quad d[u_3(0)] = \mathbf{T}(0)\mathbf{U}_3 = \Gamma_3. \tag{10.24}$$

Now, we replace the matrices Equation 10.24 in the last rows of the matrix system which is defined in Equation 10.22 [18]. Then, we get a new system as

$$\mathbf{P}_i\tilde{\mathbf{U}}_i - \mathbf{R}_i\tilde{\overline{\mathbf{U}}}_{ij} - \mathbf{S}_i\tilde{\overline{\mathbf{U}}}^*_{ijk} = \tilde{\mathbf{Z}}_i, \quad i,j,k = 1,2,3. \tag{10.25}$$

This system has $3 \times (N+1)$ numbers of nonlinear algebraic equations which is of also the initial conditions [15]. Besides it includes the unknown coefficients given as $u_{in}(t)$ for $i = 1,2,3$ and $a \leq t \leq b$ [15]. The system is solvable by using the Gaussian Elimination and has efficient results. The accuracy of the method is investigated in the following section.

10.3.3 Convergence and error bounds

In this subsection, we present a convergence analysis for the method that includes the theoretical results based on the studies in the field (see [29, 11]). Besides error bounds are given in order to show the approximation clearly.

Definition: For all $t \in [a,b]$ the approximate solution of the problem 10.1–10.4 exists and

$$E_{1,N}(t) = \left| u'_{1,N} - \phi u_{1,N}(1-u_{1,N}) + \mu u_{1,N}u_{2,N} \right|, \tag{10.26}$$
$$E_{2,N}(t) = \left| u'_{2,N} - (\eta u_{1,N} + u_{3,N})u_{2,N}(1-u_{2,N}) + u_{2,N} \right|, \tag{10.27}$$
$$E_{3,N}(t) = \left| u'_{3,N} - \zeta f(u_{2,N}) + \xi u_{3,N} \right|, \tag{10.28}$$

where $E_{1,N}$, $E_{2,N}$ and $E_{3,N}$ are called as "*absolute error functions*". In particular, $E_{i,N} \cong 0$ $(i = 1,2,3)$ are denoted as the absolute error functions of the each unknowns, respectively. Now, we employ the residual error concept to obtain an error estimation. Then, we define the following arguments.

Definition: For all $t \in [a,b]$ approximation of the problem Equation 10.1–10.4 is found and we describe the absolute error functions to check the accuracy. Moreover,

$$-R_1(t) = e'_{1,N} - \phi e_{1,N}(1 - e_{1,N}) + \mu e_{1,N} e_{2,N}, \tag{10.29}$$

$$-R_2(t) = e'_{2,N} - (\eta e_{1,N} + e_{3,N}) e_{2,N}(1 - e_{2,N}) + e_{2,N}, \tag{10.30}$$

$$-R_3(t) = e'_{3,N} - \zeta f(e_{2,N}) + \xi e_{3,N}, \tag{10.31}$$

in which $e_{1,N}$, $e_{2,N}$ and $e_{3,N}$ are called as "*error functions*" for each solutions, respectively (see [22, 20]). Furthermore,

$$R_1(t) = u'_{1,N} - \phi u_{1,N}(1 - u_{1,N}) + \mu u_{1,N} u_{2,N}, \tag{10.32}$$

$$R_2(t) = u'_{2,N} - (\eta u_{1,N} + u_{3,N}) u_{2,N}(1 - u_{2,N}) + u_{2,N}, \tag{10.33}$$

$$R_3(t) = u'_{3,N} - \zeta f(u_{2,N}) + \xi u_{3,N}, \tag{10.34}$$

where $R_i(t)$ for $i = 1,2,3$ are called "*residual functions*" ([28, 36]). So, we consider an error problem together with $u_i(0) = 0$ $(i = 1,2,3)$, i.e., the homogeneous conditions for the same procedure for m $(m > N)$ as the truncation number. Later, we have the solution $e_{i,N,m}(t)$ which satisfies the corrected polynomial solution as $u_{i,N,m}(t) = u_{i,N}(t) + e_{i,N,m}(t)$ as well as the corrected error function $e^*_{i,N,m}(t) = e_{i,N}(t) - e_{i,N,m}(t)$ for $i = 1,2,3$ in order to provide an improved results for the approximate solutions.

Corollary: Assume that $u_{i,N}(t)$ $(i = 1,2,3)$ are the numerical solutions of the problem Equation 10.1-10.4 obtained by the Taylor collocation method. Then,

$$\|e_{i,N}(t) - e^*_{i,N,m}(t)\| \le \|u_i(t) - u_{i,N}(t)\|, \quad i = 1,2,3. \tag{10.35}$$

Corollary: Assume that $e_{i,N,m}(t)$ $(i = 1,2,3)$ is the solution of the error problem Equation 10.32–10.34 with the homogeneous conditions. So, $e_{i,N,m}(t) \to e_{i,N}(t)$ converges.

Theorem 10.1

Assume that $u_i(t)$ for $i = 1,2,3$ are well-defined and differentiable functions, $t \in [a,b]$ and $u_{i,N}(t)$ for $i = 1,2,3$ are the approximate solutions of the problem 10.1–10.4 with the initial conditions, respectively. According to the mean square error,

$$\|u_i(t) - u_{i,N}(t)\|_2 \equiv \left[(u_i(t) - u_{i,N}(t))^2 \right]^{1/2}, \quad i = 1,2,3, \tag{10.36}$$

$$0 \leq \sum_{n=1}^{N} M_{n,i} |u_i - u_{i,n}|, \quad i = 1,2,3, \tag{10.37}$$

and

$$\|u_i(t) - u_{i,N}(t)\|_2 \equiv \max_{a \leq t \leq b} \|u_i(t) - u_{i,N}(t)\|, \quad i = 1,2,3. \tag{10.38}$$

Additionally, the mean square error is bounded by

$$\|u_i(t) - u_{i,N}(t)\|_2 \equiv \|G_n(t)\|_2 \left\{ \|R_{n,i}(t)\|_2 + \frac{S_n^{1/2}(t) \left\{ \sum_{n=1}^{N} \|R_{n,i}(t)\|_2^2 \|G_n(t)\|_2^2 \right\}^{1/2}}{1 - \left\{ \sum_{n=1}^{N} \|G_n(t)\|_2^2 S_n(t) \right\}^{1/2}} \right\}, \tag{10.39}$$

where $S_n(t) \equiv \sum_{n=1}^{N} M_n^2(t) G_n(t)$ *is the Green's function for the Laplacian operator of the Equations 10.1–10.4,* $M_n(t)$ *are the Lipschitz constants,* $R_{n,i}(t)$ $(i = 1,2,3)$ *denotes the residuals and we assume that the nonlinear functions in Equation 10.1-10.4 are Lipschitz continuous [11].*

Proof 10.1 First, we consider Equations 10.36, 10.29–10.34 and the Green's function for the Laplacian operator together with the homogeneous conditions of the error problem. Then, we have

$$\|u_i(t) - u_{i,N}(t)\|_2^2 = \left[G_n(t) \left(R_{n,i}(t) + S_n(t)\right)\right]^2, \quad i = 1,2,3.$$

In the next step, we apply Cauchy-Schwartz's inequality and we have

$$\begin{aligned} \|u_i(t) - u_{i,N}(t)\|_2^2 &\leq (G_n(t))^2 \left[R_{n,i}(t) + S_n(t)\right]^2, \\ \|e_{i,N}(t)\|_2 &\equiv \left[(u_i(t) - u_{i,N}(t))^2\right]^{1/2} \leq \left[(G_n(t))^2\right]^{1/2} \left[(R_{n,i}(t) + S_n(t))^2\right]^{1/2}. \end{aligned}$$

Accordingly, we apply Minkowski's inequality and

$$\|e_{i,N}(t)\|_2 \leq \|(G_n(t))\|_2 \left[\left((R_{n,i}(t))^2 \right)^{1/2} + \left((S_n(t))^2 \right)^{1/2} \right].$$

Due to the Lipschitz condition in Equation 10.39, we now consider the inequality bounded concerning the errors.

$$\begin{aligned} \|e_{i,N}(t)\|_2 &\leq \|G_n(t)\|_2 \left[\|R_{n,i}(t)\|_2 + \left(\sum_{n=1}^{N} (M_n(t))^2 \right)^{1/2} \left(\sum_{n=1}^{N} (e_{i,N}(t))^2 \right)^{1/2} \right], \\ &\leq \|G_n(t)\|_2 \left[\|R_{n,i}(t)\|_2 + \left(\sum_{n=1}^{N} (M_n(t))^2 \right)^{1/2} \left(\sum_{n=1}^{N} \|e_{i,N}(t)\|_2^2 \right)^{1/2} \right]. \end{aligned} \tag{10.40}$$

From the Mean Value Theorem

$$
\begin{aligned}
\left(\sum_{n=1}^{N}\|e_{i,N}(t)\|_2^2\right)^{1/2} &\leq \left(\sum_{n=1}^{N}\|G_n(t)\|_2^2\,(\|R_{n,i}(t)\|_2\right.\\
&+ \left.(S_n(t))^{1/2}\left(\sum_{n=1}^{N}\|e_{i,N}(t)\|_2^2\right)^{1/2}\right)^2\Bigg)^{1/2},\\
&\leq \left(\sum_{n=1}^{N}\|G_n(t)\|_2^2\|R_{n,i}(t)\|_2^2\right)^{1/2}\\
&+ \left(\sum_{n=1}^{N}\|e_{i,N}(t)\|_2^2\right)^{1/2}\left(\sum_{n=1}^{N}\|G_n(t)\|_2^2 S_n(t)\right)^{1/2},
\end{aligned}
$$

with the solution for $\left(\sum_{n=1}^{N}\|e_{i,N}(t)\|_2^2\right)^{1/2}$. Consequently, we reduce Equation 10.40 to Equation 10.39 which concludes the proof.

10.3.4 The algorithm

Input: Fixed parameters are chosen as $\phi, \eta, \zeta, \xi, \theta, \psi$ and μ. Thereby, the outputs are listed as approximation results, $u_{i,N}(t)$, $i = 1,2,3$. Hence, the steps of the algorithm can be presented as below.

Step 1. Choose truncation limit $N \in \mathbb{N}$.
Step 2. Construct the matrices with the chosen N.
Step 3. Replace in the fundamental matrix equation.
Step 4. Apply t_l the collocation points in *Step 3* and get the system.
Step 5. Repeat *Step 1-4* for the initial conditions.
Step 6. Renew the system in *Step 4*.
Step 7. Solve the system in *Step 6*.
Step 8. Check the accuracy and bounded error.
Step 9. **If** $E_{i,N} \cong 0$ $(i = 1,2,3)$ then stop, **Else** back *Step 1*.

10.4 Stability Analysis

Linear stability analysis is an important concept to investigate the local stability of an equilibrium. The basic idea is to analyse the dynamics by adding a small disturbance to the fixed point and to explore its stability characteristics. In this section, motivated by the work of [24], we perform linearisation around the equilibria of the model Equation 10.1–10.3 and discuss the dynamics of the complex interactions between pathogen and two inflammatory feedbacks.

10.4.1 Equilibrium of the model

The equilibria of the system Equation 10.1–10.3 can be found by setting

$$\frac{\mathrm{d}u_i}{\mathrm{d}t}=0, \quad i=1,2,3, \tag{10.41}$$

leading to

$$u_1\left(u_1-1+\frac{\mu}{\phi}u_2\right) = 0, \tag{10.42}$$

$$u_2\left(u_2-1+\frac{1}{\eta u_1+u_3}\right) = 0, \tag{10.43}$$

$$u_3-\frac{1}{\kappa_2}f(u_2) = 0, \quad \text{where} \quad \kappa_2=\frac{\xi}{\zeta}. \tag{10.44}$$

Substituting Equation 10.44 into 10.43 gives

$$u_2\left[u_1-\frac{1}{\eta}\left(\frac{1}{1-u_2}-\frac{1}{\kappa_2}f(u_2)\right)\right]=0, \tag{10.45}$$

and thus, the equilibria of the system can be found by solving Equations 10.42 and 10.45 together, that are

- $E_1=(0,0,\alpha)$, where $\alpha=f(0)/\kappa_2$,
- $E_2=(1,0,\alpha)$,
- $E_{3,4}=\left(0,\beta^{(j)},\kappa_2 f(\beta^{(j)})\right)$, $j=1,2$ where $\beta^{(1)}$ and $\beta^{(2)}$ are obtained by solving
$$f(u_2)(1-u_2)-\kappa_2=0,$$
- $E_5=(1-\kappa_1\gamma,\gamma,f(\gamma)/\kappa_2)$ for which Equations 10.42–10.44 can be written in terms of a single variable (let $\gamma=u_2$) as
$$\eta(1-\kappa_1\gamma)(1-\gamma)+\frac{1}{\kappa_2}(1-\gamma)f(\gamma)-1=0.$$

10.4.2 Linearisation

The local stability of the model can be determined by using linearisation around the equilibria, for which substituting the statements

$$u_i\to u_i^*+\varepsilon\tilde{u}_i, \qquad \varepsilon\ll 1, \quad i=1,2,3.$$

In the main system, the linearised version of the model is obtained for stability as follows

$$\frac{d\tilde{u}_1}{dt} = (\phi(1-2u_1^*) - \mu u_2^*)\,\tilde{u}_1 - \mu u_1^* \tilde{u}_2, \tag{10.46}$$

$$\frac{d\tilde{u}_2}{dt} = \eta u_2^*(1-u_2^*)\tilde{u}_1 + [(\eta u_1^* + u_3^*)(1-2u_2^*) - 1]\,\tilde{u}_2 + u_2^*(1-u_2^*)\tilde{u}_3, \tag{10.47}$$

$$\frac{d\tilde{u}_3}{dt} = \frac{\zeta}{\psi}\operatorname{sech}^2\left(\frac{u_2^* - \theta}{\psi}\right)\tilde{u}_2 - \xi \tilde{u}_3. \tag{10.48}$$

Here $E^* = (u_1^*, u_2^*, u_3^*)$ denotes a generic expression for the equilibria and the variables with $\tilde{\cdot}$ represent perturbed quantities. Accordingly, this can be rewritten as

$$\frac{d}{dt}\begin{pmatrix} u_1 \\ u_2 \\ u_3 \end{pmatrix} = \mathscr{J}|_{E^*}\begin{pmatrix} u_1 \\ u_2 \\ u_3 \end{pmatrix}, \tag{10.49}$$

where

$$\mathscr{J}|_{E^*} = \begin{pmatrix} \phi(1-2u_1^*) - \mu u_2^* & -\mu u_1^* & 0 \\ \eta u_2^*(1-u_2^*) & (\eta u_1^* + u_3^*)(1-2u_2^*) - 1 & u_2^*(1-u_2^*) \\ 0 & \frac{\zeta}{\psi}\operatorname{sech}^2\left(\frac{u_2^*-\theta}{\psi}\right) & -\xi \end{pmatrix}. \tag{10.50}$$

Here $\tilde{\cdot}$ are omitted for simplicity. Note that the Jacobian matrix given in Equation 10.50 is generic, thus, the stability of each steady state can be easily computed. For example the eigenvalues of the equilibrium $E_1 = (0, 0, \alpha)$, $\alpha = f(0)/\kappa_2$ can be obtained by using $\det(\mathscr{J}|_{E_1} - \lambda I)$, where I is a (3×3)-dimensional characteristic matrix. Therefore, the characteristic equation is found as follows

$$(\phi - \lambda)(\alpha - 1 - \lambda)(-\xi - \lambda) = 0, \tag{10.51}$$

leading to eigenvalues of $\lambda_1 = \phi$, $\lambda_2 = \alpha - 1$ and $\lambda_3 = -\xi$. It is straightforward to see that $\lambda_1 = \phi > 0$ and thus the steady state $E_1 = (0, 0, \alpha)$ is always unstable. The eigenvalue analysis for other equilibria (e.g., for E_i, $i = 2, 3, 4, 5$) can be performed in a similar manner.

10.5 Numerical Simulations

In this section, we perform some numerical simulations for the system Equation 10.1–10.3 including the dynamics of pathogen and two inflammatory mediators [24]. Solutions of this model can either be oscillatory or converge to its steady state, depending upon the choice of parameter set and initial conditions. As an illustrative example, we particularly focus on the parameter continuation for parameters μ, ξ and η. Numerical results for time evolution of the components are performed using MATLAB® 2017 and bifurcation analyses are preformed using

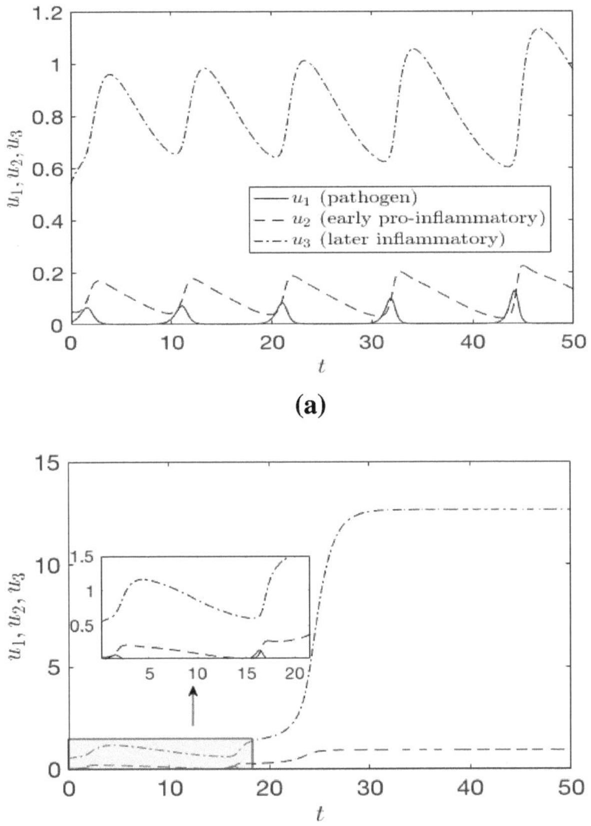

(b)

Figure 10.1: Time evolution of the variables u_1, u_2 and u_3 are shown for two different levels of μ, where $\mu = 30$ (a) and $\mu = 25$ (b). The initial conditions are given as $u_1(0) = 0.01$, $u_2(0) = 0.05$, $u_3(0) = 0.539$. Other parameters are $\phi = 3, \eta = 15, \zeta = 15, \xi = 1, \theta = 1, \psi = 0.5$.

DDE-BIFTOOL [10]. Besides, the present technique is applied and accuracy is shown by tables and figures with the comparison of the Runge-Kutta method with fourth order (RK) and the Taylor collocation approach with the Multi step algorithm (TMS) by using MATLAB and Maple softwares (cf. [26, 30, 32, 33]).

In Figure 10.1, unstable oscillatory (a) and stable (b) dynamics of pathogen (u_1), early pro-inflammatory (u_2) and later inflammatory (u_3) mediators are shown, for $\mu = 30$ (a) and $\mu = 25$ (b), respectively. Decreasing $\mu = 30$ to $\mu = 25$, in Figure 10.1(b), the pathogen is cleared from the system at $t = 17$ and pro-inflammatory (u_2) and later inflammatory (u_3) mediators return to their resting state at $u_2 = 0.9209$ and $u_3 = 12.65$, respectively.

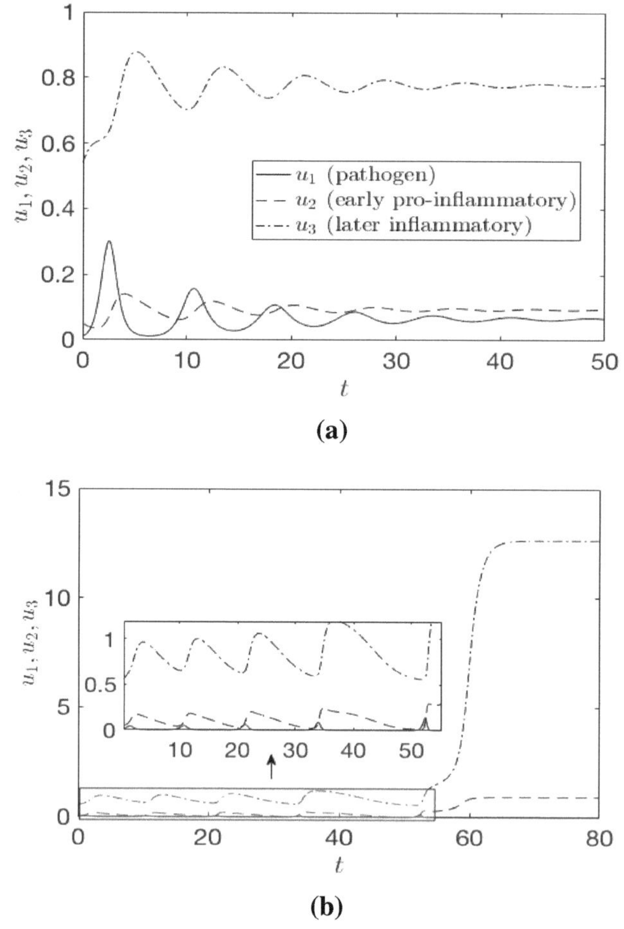

Figure 10.2: Time evolution of the variables u_1, u_2 and u_3 are shown for two different levels of η, where $\eta = 5$ (a) and $\eta = 40$ (b). Initial conditions are chosen same as in Figure 10.1. Other parameters are $\phi = 3, \mu = 30, \zeta = 15, \xi = 1, \theta = 1, \psi = 0.5$.

Time evolution of the pathogen (u_1), two inflammatory mediators (u_2 and u_3) are presented in Figure 10.2 for $\eta = 5$ (a) and $\eta = 40$ (b). Damping oscillations for all components are observed, leading to a stable spiral in Figure 10.2(b). Increasing η significantly, the system converges to equilibria E_3 or E_4 and eventually, becomes stationary after displaying some oscillations for a finite time, as seen in Figure 10.2(b).

The dynamic of the model can be alternatively presented by using three-dimensional trajectories for phase diagrams as shown in Figure 10.3, where a stable spiral for $\eta = 5$ converging to positive coexisting state (a) and unstable positive equilibrium surrounded with a stable limit cycle (b) can be observed.

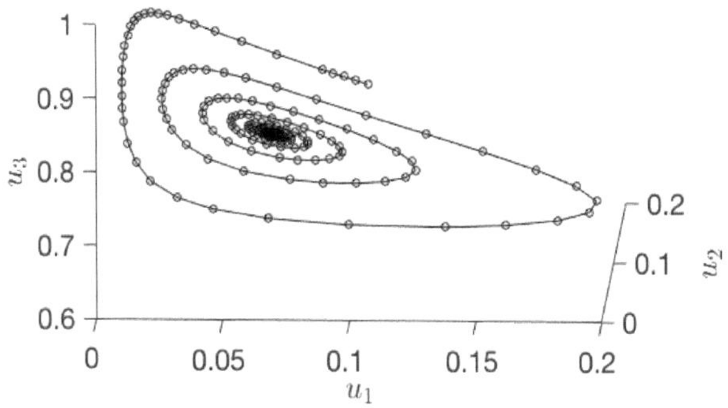

(a)

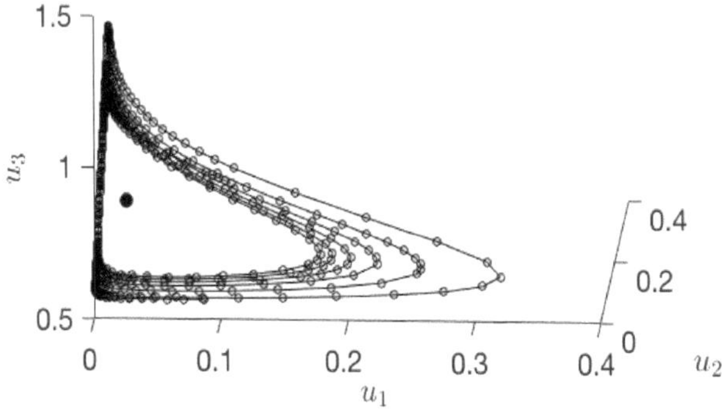

(b)

Figure 10.3: The phase diagrams corresponding to dynamics for $\eta = 5$ (a) and $\eta = 15$ (b) are shown. The initial conditions are chosen as $u_1(0) = 0.1$, $u_2(0) = 0.15$ and $u_3(0) = 0.8$. The equilibrium points for each case are shown with black dot.

Figures 10.4 and 10.5 demonstrate a numerical continuation analysis to complement the theoretical study given in Sections 10.4.1 and 10.4.2. Here, two inflammatory responses (u_2 and u_3) are examined under parameter variation and stability of the steady states presented in Section 10.4.1 is demonstrated. Along the solution branches, several type of bifurcations including Hopf bifurcation (HB), saddle node bifurcation (SNB) and transcritical bifurcation (TCB) are detected. On the other hand, $\lambda_N = k$, $k = 0, 1, 2, 3$ stands for the number of eigenvalues with a positive real part. In the graph, the solid line indicates stability, the dashed line represents instability with one eigenvalue with a positive real part, and dotted and dash-dot lines, respectively, describe eigenvalues existing with two and three positive real parts. For instance, changes in early pro-inflammatory (u_2) and later inflammatory (u_3) dynamics for parameters μ and ξ are shown in Figures 10.4(a, b) and 10.5(a, b), where u_2 and u_3 demonstrate similar stability behaviour. Here, the pathogen free state (E_5) for both inflammatory responses encounters a stability change if μ is under a critical value, whilst an unstable limit

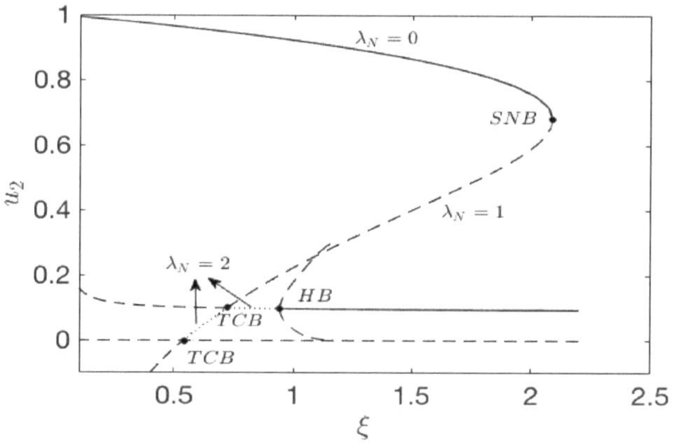

(a)

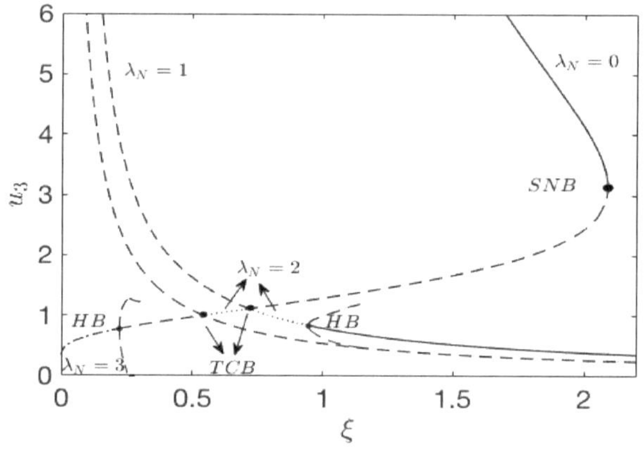

(b)

Figure 10.4: Bifurcation diagrams of u_2 (early pro-inflammatory response) and u_3 (later inflammatory mediator) with respect to parameter μ. Here solid line shows stable dynamics, dashed line shows unstable dynamics with one eigenvalue with positive real part ($\lambda_N = 1$), and dotted line shows dynamics with two eigenvalues with positive real part ($\lambda_N = 2$) and thus unstable. In both (a) and (b), there is one periodic orbit arising from Hopf bifurcation HB. Other parameters are $\phi = 3, \xi = 1, \eta = 15, \zeta = 15$.

cycle appears through subcritical Hopf bifurcation when μ exceeds its critical value, that is $\mu = 26.59$ for both cases. Note that $E_3 = (0, 0.9209, 12.6476)$ is stable when crossing $\mu > 3.258$ and $E_4 = (0, 0.2241, 1.2889)$ changes its stability from $\lambda_N = 2$ to $\lambda_N = 1$ at $\mu = 13.39$, both occur via transcritical bifurcation. Furthermore, E_2 is always unstable regardless of the parameter μ.

In Figure 10.5, the bifurcation diagrams of both inflammatory response are shown with respect to parameter ξ which stands for the natural depletion of a

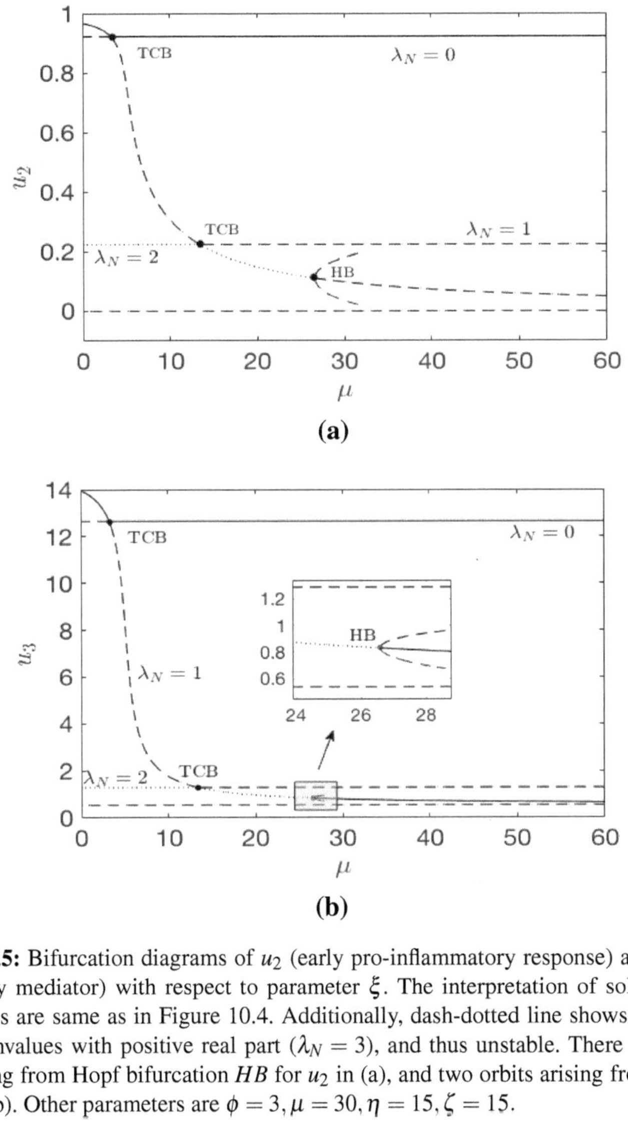

Figure 10.5: Bifurcation diagrams of u_2 (early pro-inflammatory response) and u_3 (later inflammatory mediator) with respect to parameter ξ. The interpretation of solid, dashed and dotted lines are same as in Figure 10.4. Additionally, dash-dotted line shows dynamics with three eigenvalues with positive real part ($\lambda_N = 3$), and thus unstable. There is one periodic orbit arising from Hopf bifurcation HB for u_2 in (a), and two orbits arising from Hopf points for u_3 in (b). Other parameters are $\phi = 3, \mu = 30, \eta = 15, \zeta = 15$.

later inflammatory mediator. Similar to Figure 10.4, transcritical bifurcation occurs at the intersection of steady states. In both cases, a subcritical Hopf bifurcation is detected at $\xi = 0.9467$. Besides, an additional subcritical Hopf bifurcation can be observed for u_3, where λ_N is decreased from 3 to 1. The equilibria E_3 and E_4 move towards each other and collide in a saddle node bifurcation at $\xi = 2.088$, in which the upper branch is stable and the lower branch is unstable.

Now, we consider the error comparison of the numerical solutions in Table 10.1 and Table 10.2. In Table 10.1, it can be seen easily that we have sufficient results whenever N the truncation limit and m the residual error truncation are chosen large enough. This generalization has been already mentioned in some studies which is also proven in our study ([15, 19, 22, 20]). Here, we can also give a specific comment on the residual corrected errors with more efficient results as can can be seen in Table 10.1.

Table 10.1: Error convergence for different N, m values.

t	$e_{4,5}(t)$	$e_{6,7}(t)$	$e_{8,9}(t)$
0.0	0.22356E-3	0.30001E-4	0.25097E-5
0.5	0.62484E-2	0.01914E-3	0.79235E-4
1.0	0.84023E-2	0.41259E-3	0.52082E-4

On the other hand, in Table 10.2 we can observe the comparison of the errors between two numerical solutions for the same truncation error $N = 4$. From here, we can show that TMS solutions provide more accurate results than RK numerical solutions. In this table we can also have an idea about the residual error investigation which gives more efficient results whenever it is involved in the solution. This result is visible with the help of the comparison of Table 10.1 and Table 10.2 between the errors $e_{4,5}(t)$ and $e_4(t)$, respectively.

Table 10.2: Error comparison of RK and TMS for $N = 4$.

t	$e_4(t)$ **in RK**	$e_4(t)$ **in TMS**
0.0	0.92630E-3	0.10352E-2
0.5	0.83524E-2	0.85953E-2
1.0	0.95684E-2	0.92381E-2

In Figure 10.6, we can see the RK solutions in the fourth order for different μ values. We can also see the comparison of the solutions for TMS and RK numerical solutions in Figure 10.6 for $N = 4$ and with the fourth order, respectively. In these findings, it is seen that the efficient approximation results are obtained from the present TMS technique.

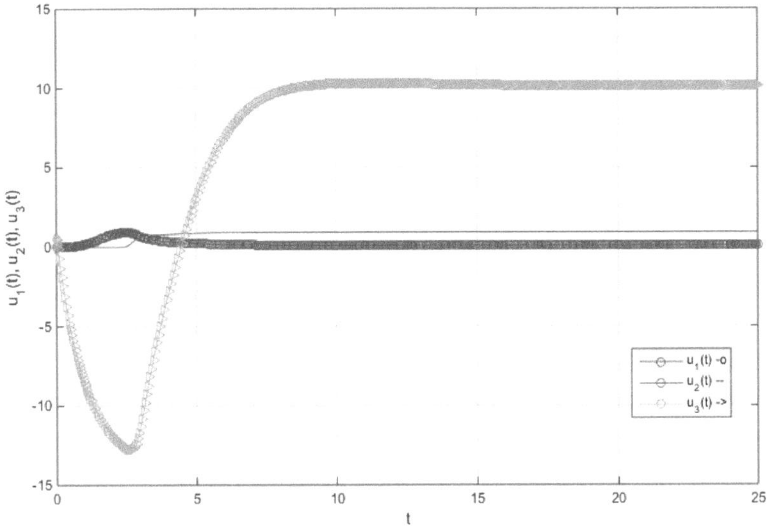

(a)

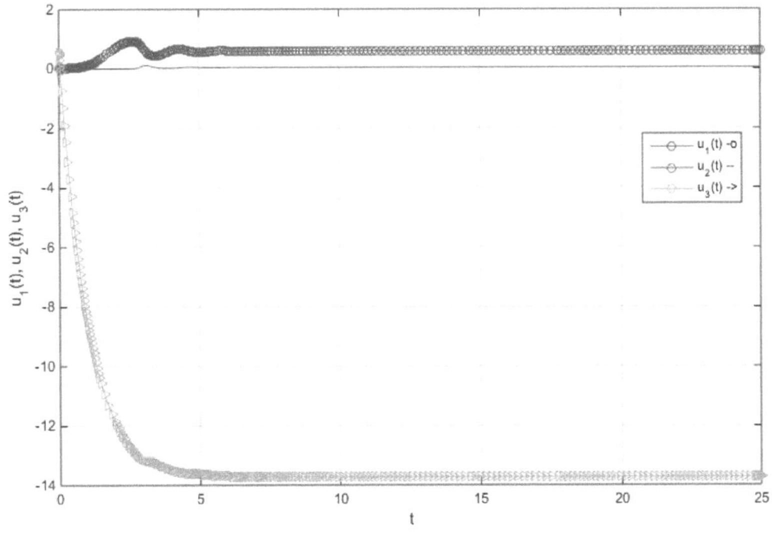

(b)

Figure 10.6: Numerical solution of the with RK fourth order solution where (a) shows the result for $\mu = 3$ and (b) shows the result for $\mu = 30$.

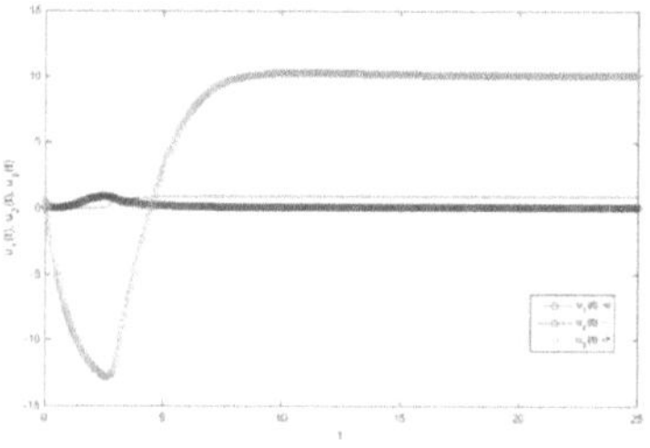

(a)

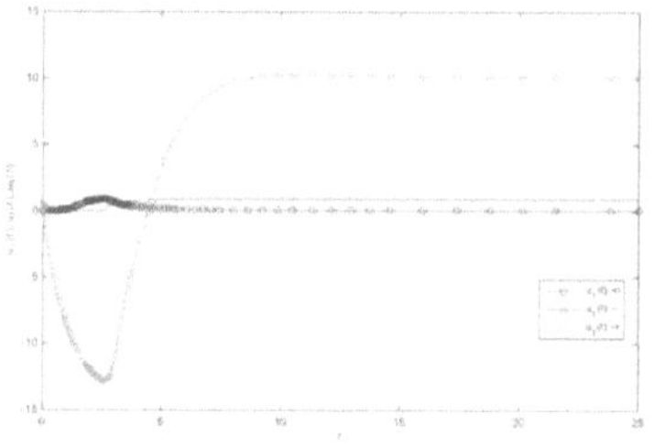

(b)

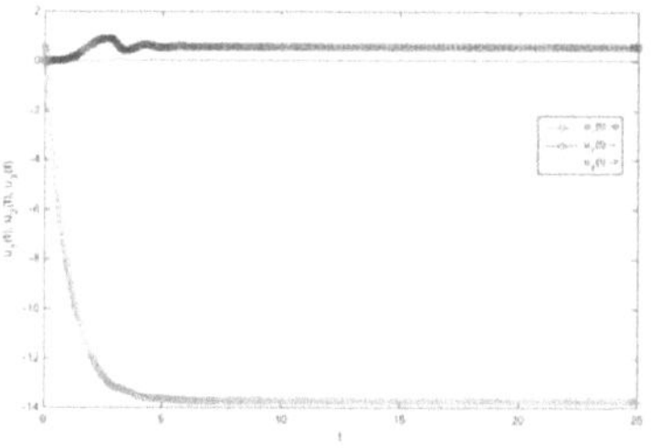

(c)

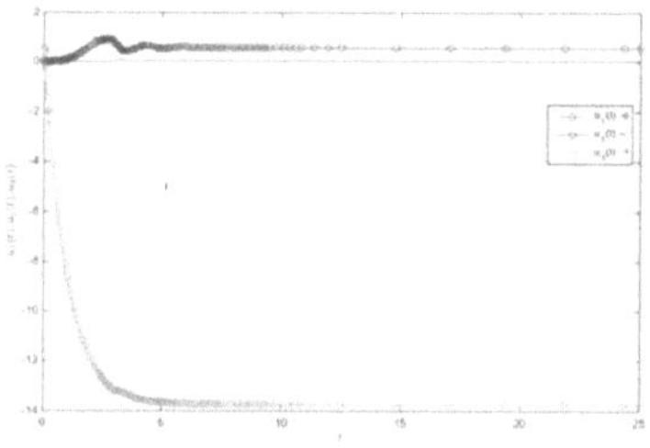

(d)

Figure 10.7: Numerical solution comparison between RK fourth order and TMS for $N = 4$. Here, (a) and (b) show the results for $\mu = 3$, RK solution and TMS solution, respectively. Besides (c) and (d) show the results for $\mu = 30$, RK solution and TMS solution, respectively.

10.6 Conclusion and Outlook

In this chapter, various numerical and theoretical investigations of a mathematical model of acute inflammation are presented. Firstly, the solutions of the model are provided by using a numerical method, where the model equations are expressed via the matrix equations. Thus, a pointwise convergence for the solutions can be obtained by implementing the collocations points. Secondly, the stability of the model is discussed to better understand the interactions between a pathogen and two pro-inflammatory responses and their dynamical properties.

There are several extensions of the work which we have presented here to treat more realistic biological models of acute inflammation; we can outline some of them: Since the body's inflammatory response to pathogen requires some time lag, one possible extension would be to incorporate delay terms in the model [13]. Furthermore, in temporal models, organisms are assumed to be homogeneously distributed. It is well recognised that spatio-temporal models of interacting species are known to provide an extension to temporal models. Thus, extending the model with the non-local mobility of organisms, leading to spatio-temporal model, would be another natural extension [12]. Moreover, one obvious caveat of the model Equation 10.1–10.3 is that the pathogen can never be eliminated from the system in finite time (see Figure 10.1(a)). However, in reality, one expects that an inflammatory response should eliminate the pathogen and then, get back to the resting state, which is not the case here. Therefore, as stated in [24], it is more reasonable to set a threshold, under which the pathogen is cleared from the system. Numerical analysis of this case is reserved for the future. Additionally, the error estimation of this study together with our findings give an original contribution to this field. On the other hand, systems of ordinary differential equations are very important in mathematical biological models in biology and medical information theory (see [3, 8, 14, 38]). Moreover, uncertainty modeling of dynamical systems is also important in the field and recent applications exist ([23, 25]). Due to this reason, the method and our strategy in this study can be modified for such systems as another future outlook of this study.

References

[1] K. Abdella and J. Trivedi. Solving multi-point boundary value problems using sinc-derivative interpolation. *Mathematics*, 8(12): 2104, 2020.

[2] C. Arévalo, C. Führer and M. Selva. A collocation formulation of multistep methods for variable step-size extensions. *Applied Numerical Mathematics*, 42(1-3): 5–16, 2002.

[3] E. Ayyıldız, V. Purutçuoğlu and G.W. Weber. Loop-based conic multivariate adaptive regression splines is a novel method for advanced construction of complex biological networks. *European Journal of Operational Research*, 270(3): 852–861, 2018.

[4] G. Bontempi and Y.Le Borgne. An adaptive modular approach to the mining of sensor network data. In *Proceedings of the Workshop on Data Mining in Sensor Networks, SIAM SDM*, pp. 3–9. SIAM Press, 2005.

[5] M. Bosmann and P.A. Ward. The inflammatory response in sepsis. *Trends in Immunology*, 34(3): 129–136, 2013.

[6] J.S. Calvert, D.A. Price, U.K. Chettipally, C.W. Barton, M.D. Feldman, J.L. Hoffman, M. Jay and R. Das. A computational approach to early sepsis detection. *Computers in Biology and Medicine*, 74: 69–73, 2016.

[7] A. Chowdhury, S. Clayton and M. Lemma. Numerical solutions of nonlinear ordinary differential equations by using adaptive runge-kutta method. *Journal of Advances in Mathematics*, 17(12): 147–154, 2019.

[8] S. Daun, J. Rubin, Y. Vodovotz and G. Clermont. Equation-based models of dynamic biological systems. *Journal of Critical Care*, 23(4): 585–59, 2008.

[9] J. Day, J. Rubin, Y. Vodovotz, C.C. Chow, A. Reynolds and G. Clermont. A reduced mathematical model of the acute inflammatory response II. Capturing scenarios of repeated endotoxin administration. *Journal of Theoretical Biology*, 242(1): 237–256, 2006.

[10] K. Engelborghs, T. Luzyanina and G. Samaey. Dde-biftool: A matlab package for bifurcation analysis of delay differential equations. *TW Report*, 305: 1–36, 2000.

[11] N.B. Ferguson and B.A. Finlayson. Error bounds for approximate solutions to nonlinear ordinary differential equations. *AIChE Journal*, 18(5): 1053–1059, 1972.

[12] A. Gökçe, D. Avitabile and S. Coombes. The dynamics of neural fields on bounded domains: An interface approach for dirichlet boundary conditions. *The Journal of Mathematical Neuroscience*, 7(1): 1–23, 2017.

[13] A. Gökçe, S. Yazar and Y. Sekerci. Delay induced nonlinear dynamics of oxygen-plankton interactions. *Chaos, Solitons & Fractals*, 141: 110327, 2020.

[14] N. Gökgöz, H. Öktem and G.W. Weber. Modeling of tumor-immune nonlinear stochastic dynamics with hybrid systems with memory approach. *Results in Nonlinear Analysis*, 3(1): 353–377, 2020.

[15] E. Gökmen, O.R. Işık and M. Sezer. Taylor collocation approach for delayed Lotka-Volterra predator-prey system. *Applied Mathematics and Computation*, 268(3): 671–684, 2015.

[16] B. Gürbüz. Laguerre matrix-collocation technique to solve systems of functional differential equations with variable delays. *AIP Conference Proceedings*, 2183(1): 090007, 2019.

[17] B. Gürbüz. A computational approximation for the solution of retarded functional differential equations and their applications to science and engineering. *arXiv preprint arXiv:2103.09046*, 2021.

[18] B. Gürbüz. A numerical scheme for the solution of neutral integro-differential equations including variable delay. *Mathematical Sciences*, pp. 1–9, 2021.

[19] B. Gürbüz, H. Mawengkang, I. Husein and G.W. Weber. Rumour propagation: An operational research approach by computational and information theory. *Central European Journal of Operations Research*, pp. 1–21, 2021.

[20] B. Gürbüz and M. Sezer. An hybrid numerical algorithm with error estimation for a class of functional integro-differential equations. *Gazi University Journal of Science*, 29(2): 419–434, 2016.

[21] B. Gürbüz and M. Sezer. Laguerre matrix-collocation method to solve systems of pantograph type delay differential equations. *International Conference on Computational Mathematics and Engineering Sciences*, pp. 121–132, 2019.

[22] B. Gürbüz, M. Sezer, and C. Güler. Laguerre collocation method for solving Fredholm integro-differential equations with functional arguments. *Journal of Applied Mathematics*, 2014: 1–13, 2014.

[23] E. Kropat, G.W. Weber and B. Akteke-Öztürk. Eco-finance networks under uncertainty. In *Proceedings of the International Conference on Engineering Optimization*, pp. 353–377, 2008.

[24] R. Kumar, G. Clermont, Y. Vodovotz and C.C. Chow. The dynamics of acute inflammation. *Journal of Theoretical Biology*, 230(2): 145–155, 2004.

[25] Z. Liu and X. Yang. A linear uncertain pharmacokinetic model driven by Liu process. *Applied Mathematical Modelling*, pp. 1881–1899, 2021.

[26] Maplesoft. A division of Waterloo Maple Inc. Waterloo, Ontario. *https://www.maplesoft.com/*, 2018.

[27] C. Nedeva, J. Menassa and H. Puthalakath. Sepsis: Inflammation is a necessary evil. *Frontiers in Cell and Developmental Biology*, 7: 108, 2019.

[28] F.A. Oliveira. Collocation and residual correction. *Numerische Mathematik*, 36(1): 27–31, 1980.

[29] J. Oliver. An error estimation technique for the solution of ordinary differential equations in Chebyshev series. *The Computer Journal*, 12(1): 57–62, 1969.

[30] MATLAB R2014b. The MathWorks, Natick. *https://de.mathworks.com/*, 2014.

[31] A. Reynolds, J. Rubin, G. Clermont, J. Day, Y. Vodovotz and G.B. Ermentrout. A reduced mathematical model of the acute inflammatory response: I. Derivation of model and analysis of anti-inflammation. *Journal of Theoretical Biology*, 242(1): 220–236, 2006.

[32] L.F. Shampine and M.W. Reichelt. The matlab ode suite. *SIAM Journal on Scientific Computing*, 18(1): 1–22, 1997.

[33] L.F. Shampine, M.W. Reichelt and J.A. Kierzenka. Solving index-1 DAEs in MATLAB and Simulink. *SIAM Journal on Scientific Computing*, 41(3): 538–552, 1999.

[34] N. Shawagfeh and D. Kaya. Comparing numerical methods for the solutions of systems of ordinary differential equations. *Applied Mathematics Letters*, 17(3): 323–328, 2004.

[35] R. Tirani and C. Paracelli. An algorithm for starting multistep methods. *Computers & Mathematics with Applications*, 45(1-3): 123–129, 2003.

[36] K. Wang and Q. Wang. Taylor collocation method and convergence analysis for the Volterra-Fredholm integral equations. *Journal of Computational and Applied Mathematics*, 260: 294–300, 2014.

[37] W. Wasow. Solution of certain nonlinear differential equations by series of exponential functions. *Illinois Journal of Mathematics*, 2(2): 254–260, 1958.

[38] G.W. Weber, A. Tezel, P. Taylan, A. Soyler and M. Çetin. Mathematical contributions to dynamics and optimization of gene-environment networks. *Optimization*, 57(2): 353–377, 2008.

[39] Ş. Yüzbaşı and M. Sezer. An exponential matrix method for numerical solutions of Hantavirus infection model. *Applications & Applied Mathematics*, 8(1): 99–115, 2013.

[40] Ş. Yüzbaşı and M. Sezer. An exponential approach for the system of nonlinear delay integro-differential equations describing biological species living together. *Neural Computing and Applications*, 27(3): 769–779, 2016.

Chapter 11

Bayesian Inference for Undirected Network Models

Hajar Farnoudkia[1] and *Vilda Purutçuoğlu*[2,*]

11.1 Introduction

In almost every field including biology and industrial engineering, when the relationship between the factors or variables is the case, conditional dependence is one of the most important relationships that has been investigating within the last twenty years through the Bayesian approach. The conditional dependence is shown mostly by the undirected relationship between two corresponding variables that is applicable when the direction is not important. However, in some cases the direction of the relationship and also its power are important. In this kind of problem, the classical methods cannot be preferable. Here, some modifications can update them to be used in the inference of directed relations. One of the well-known and optimum ways to catch this kind of relationship is the information entropy where both direction and the value of a relationship are caught, which does not need any specific assumption.

The major challenge in conditional dependence is that, in some cases, the number of parameters resulting in the true model being unknown or even the number of parameters exceeds the sample size. To solve the underlying uncer-

[1] Faculty of Economics and Administrative Sciences, Başkent University.
Email: hajar.farnoudkia@gmail.com
[2] Middle East Technical University.
* Corresponding author: vpurutcu@metu.edu.tr

tainty, the Bayesian approach is performed, especially, when the contingency matrix is sparse. Additionally, it is accepted that prior knowledge about the parameter distribution can modify the estimation in a good way. On the other side, the Markov chain Monte Carlo (MCMC) methods are the most popular approaches to infer the associated parameter estimation.

In general, MCMC methods are the iterative algorithms that generate at each iteration, a random variable from the posteriors of the parameters. These proteins yield a Markov chain from which an estimator is computed (Mazet and Brie, 2006 [7]). In each iteration of this chain we get closer to the target estimation and the sampling from the posterior distribution is stopped when there is no significant difference between the estimated matrix in consequence iterations. In MCMC, the first specific proportion is discarded as a burn-in period because of the high fluctuation caused by the high difference between estimated matrices. That means in the burn-in period, the estimated matrix is changed over each iteration, but, after the burn-in period, it goes smoother in such a way the mean of posterior distribution after the burn-in period can be used to estimate the parameter matrix. In this study, when we discuss a matrix, we mean a matrix that composes the whole network or the matrix that makes the model of the network representing the undirected relationship between the variables in the model. These variables can be proteins in biology, some factors that affect a machine performance in industry or even the portfolio types which are financially related to each other. So, an undirected network is a broad concept that can be used in various fields.

There are several ways to overcome the problem of the model selection such as different types of MCMC methods, like Birth-and-death MCMC (BDMCMC) and Reversible Jump MCMC (RJMCMC), or some of their alternatives those are not so common these days, like split-merge and Carlin-Chib methods. But one of the most common ways to model the network without any specific restriction or assumption is the copula method. In this study, we explain the underlying MCMC approaches in detail and the rest briefly to acquaint with their assumptions and methodologies behind them.

11.2 Copula Gaussian Graphical Model (CGGM)

Suppose we have a data matrix with p variables and n samples and we are interested in obtaining the relationship (conditional dependence) between those variables. In this kind of network which is common in social surveys and biological aspects, each variable is shown by a node in the graph and the conditional dependence between two nodes is shown by an undirected edge. In application, the conditional dependence is determined based on the inverse of a covariance matrix, also called the precision matrix when the distribution of the variables is normal. Now, assume that the vector Y follows a p-dimensional multivariate normal distribution $N_p(0, \Theta^{-1})$ where Θ indicates the precision matrix. Hence, with

n samples from Y, i.e., $Y^{1:n}$), the likelihood function given Θ is proportional to

$$p(Y^{1:n} \mid \Theta) \propto |\Theta|^{n/2} exp\{-\frac{1}{2}tr(\Theta^T U)\}, \tag{11.1}$$

where $|.|$ shows the determinant of the given matrix, $(.)^T)$ implies the transpose of the term in the parentheses, and U is the trace of the $Y'Y$ matrix. In the Equation 11.1 we use the proportional notation, rather than equality, since the constant term is eliminated to see the more important items of the joint distribution function.

So, a graphical model with (V,E) with V nodes and E edges between nodes, from $N_p(0,\Theta^{-1})$ is called the Gaussian graphical model (GGM). In this model, if the normality assumption does not hold for the data matrix, we would not be able to use the precision matrix to understand the conditional structure of the network. So, we need a normally distributed data that is possible in the light of the Copula.

11.2.0.1 Gaussian copula

The Gaussian copula can combine the data in a way that their joint distribution is Gaussian with the same covariance matrix. For continuous data, the Gaussian copula is defined as a function that uses the empirical inverse of the cumulative distribution of each variable. This concept is explained in detail in Section 11.4. However, for binary and ordinal categorical data, Muthen (1984) [9] introduce a continuous latent variable Z by defining some thresholds $\tau_\upsilon = (\tau_{\upsilon,0}, \tau_{\upsilon,1}, ..., \tau_{\upsilon,\omega_\upsilon})$ with $-\infty = \tau_{\upsilon,0} < \tau_{\upsilon,1} < ... < \tau_{\upsilon,\omega_\upsilon} = \infty$. So,

$$y_\upsilon^j = \sum_{l=1}^{\omega_\upsilon} l \times 1_{\tau_{\upsilon,l-1} < z_\upsilon^j \leq \tau_{\upsilon,l}}, j = 1,2,...,n. \tag{11.2}$$

The relationship between Y_{ij} and Z_{ij} satisfie the constraint

$$y_{ij} < y_{ik} \Rightarrow z_{ij} < z_{ik}, z_{ij} < z_{ik} \Rightarrow y_{ij} \leq y_{ik}.$$

This means that sometimes the same y's can result from different latent variables, but, we are sure that different y's imply different z's. By defining a correlation matrix as

$$\Upsilon_{i,j}(\Theta) = \frac{(\Theta^{-1})_{i,j}}{\sqrt{(\Theta^{-1})_{i,i}(k^{-1})_{j,j}}} \tag{11.3}$$

and $Z_V \sim N_p(0,\Theta^{-1})$ which has a one-to-one correspondence with observed data.

So, in the application of the copula Gaussian graphical model (CGGM) for the construction of biological networks, it is found that RJMCMC has a long burn-in period and additionally, it needs to calculate the Jacobian term for each iteration associated to Equation 11.5.

In order to decrease the computational demand, Mohammadi and Wit (2015) [8] suggest the birth-and-death method. RJMCMC with the split-merge approach have been also intensively studied by the work of Green (1997) [4]. On the other hand, there are other alternatives to this method which have not been proposed yet for the biological networks. For instance, the Carlin and Chib algorithm (Carlin and Chibs 1995, [2]) and the Gibbs algorithm (Walker, 2009 [12]) are in this group. In discrete times, RJMCMC moves via several methods such as birth and death moves and the split-merge method.

In the following part, initially, we describe RJMCMC and then, explain its alternatives together with the Gibbs sampling which we first time introduce, and present our comparative analysis with its strong alternatives.

11.2.1 Reversible jump Markov chain Monte Carlo method (RJMCMC)

The Reversible Jump Markov chain Monte Carlo (RJMCMC) method is one of the well known approaches to unravel both challenges at the same time (Green, 1995, [4]). RJMCMC is the modified version of the Metropolis-Hasting method which provides jumps between spaces of different dimensions. In this algorithm, the move from matrix or model A to A' is not always possible because of the modality of Metropolis-Hasting. The acceptance probability for this movement from (k, θ^k) to $(k', \theta^{k'})$ is computed as

$$R_{k,k'} = \frac{\tilde{q}(k, \theta^k | k', \theta^{k'}) P(k', \theta^{k'} | y)}{\tilde{q}(k', \theta^{k'} | k, \theta^k) P(k, \theta^k | y)}, \tag{11.4}$$

where k is the dimension of the precision matrix θ and y refers to the normal random variables of the current position and finally, k' and θ' denote the associated proposal terms of k and θ, respectively. In Equation 11.4, P shows the likelihood function and $\tilde{q}$ presents the kernel density. If you notice the acceptance probability more carefully, it can be understood that the nominator is some kind of posterior probability based on the changed matrix and the denominator is the posterior probability based on the unchanged matrix. So, if the nominator overweights the denominator, it means the changed matrix increases the posterior probability and this update or move from k and θ to k' and θ' is possible.

Under the "dimension matching" condition, $(dim(\theta^{k'},x) = dim(\theta^k,y))$, where x and y are variables drown from the proposal distribution $\tilde{q}_1$, the acceptance probability is equivalent to

$$R_{k,k'} = \frac{P(k', \theta^{k'} | y)}{P(k, \theta^k | y)} \times \frac{\tilde{q}_1(k|k')\tilde{q}_2(x)}{\tilde{q}_1(k'|k)\tilde{q}_2(y)} \times \left| \frac{\partial(\theta^{k'},x)}{\partial(\theta^k,y)} \right|. \tag{11.5}$$

In Equation 11.5, $\tilde{q}_2(.)$ refers to the kernel for the given random variable and $\left|\frac{\partial(\theta^{k'},x)}{\partial(\theta^{k},y)}\right|$ represents the determinant of the Jacobian matrix for the transformation.

In the study of Dobra and Len (2011) [3], this method is used to infer the parameters of CGGM.

11.2.2 RJMCMC with birth-and-death moves

In this kind of RJMCMC, the change in the dimension of the model parameters, resulting in the model itself, is done by births and deaths steps which are proposed with probabilities P_B and P_D, respectively.

Given $Y_s = (k, \theta_{1:k})$, Green (1995) [4] suggests the following calculation:

- The birth step is computed with a probability P_B by drawing the graph G from the density $G \sim q$ as θ_{k+1}. So $dim(Y_s) : k \to (k+1)$,
- The death step is found with a probability P_D, by selecting one of $\theta_j, j = 1,...,k$ and dropping the corresponding parameter from $\theta_{1:k}$. So $dim(X_s) : k \to (k-1)$.

Hence, the acceptance probability for the birth move in a k-component configuration is $min(1,A)$ via

$$A = \frac{\pi(k+1,\theta^k)}{\pi(k,\theta^k)} \times \frac{P_D/k+1}{P_B q(\theta)} \tag{11.6}$$

in which π is a stationary distribution, q presents the proposal kernel. Thus, the acceptance probability for the death is found from $min(1,\frac{1}{A})$.

11.2.3 RJMCMC with split-merge moves

In this method, there is a deterministic invertible transformation function $t : R^2 \to R^2$ playing the main role as below:

- The split step is calculated with a probability P_S by initially choosing one of the indices j from $1,...,k$ and a $G \sim q$. Then, we put them into the t function $R^2 \to R^2$. The result has two components formed by the combination of j and G under t. So $dim(Y_s) : k \to (k+1)$.
- The merge is obtained with a probability P_M by initially selecting two indices i and j from $1,...,k$, and then, by putting them into the invertible t function $R^2 \to R^2$. Finally, the first component of $t^{-1}(\theta_i, \theta_j)$ is assigned in place of two parameters so that $dim(Y_s) : k \to (k-1)$.

Hence, the proposal probability density function depends on both P_S, P_M, q and the Jacobian of t.

11.3 RJMCMC Alternatives

As briefly discussed in the Introduction, there are also some alternatives approaches to RJMCMC. These are the Carlin-Chib algorithm, the birth-death MCMC (BDMCMC), a special case of the Gibbs sampling and finally, the quadratic approximation for the sparse inverse Covariance Estimation (QUIC) method which does not use the Bayesian approach. Below, we present the mathematical details of each method.

11.3.1 Birth-and-death MCMC (BDMCMC)

This method is based on the continuous time approach where the dimension of the parameter is not fixed. In this approach, new components are born according to the Poisson process with a rate λ_B and the ith component in a k-component configuration which dies with a rate

$$\lambda_D(i) = \frac{\pi(k-1, \theta_{1:i-1}, \theta_{i+1:k})}{\pi(k, \theta_{1:k})} \times \lambda_B q(\theta_i). \tag{11.7}$$

In Equation 11.7, $\pi(.)$ is the density kernel when $\theta_{1:k}$ implies the first k parameters and $q(\theta_i)$ represents the proposal kernel for the ith component of the parameter θ, as used beforehand.

The choice of the birth and the death rates determines the birth-death process and is made in such a way that the stationary distribution is precisely the posterior distribution of interest. Contrary to the RJMCMC approach, the moves between models are always accepted, which makes the BDMCMC approach extremely efficient.

11.3.2 Carlin-Chib algorithm

In the application of the MCMC technology to any problem involving a choice between k competing the Bayesian model specification, M is defined as an integer-value parameter that indexes the model collection. The Carlin-Chib algorithm shows how the Gibbs sampling methodology may be a specific method to choose across finite collections of models without destroying the convergence.

Suppose that $f(y|\theta_j, M = j)$ is the corresponding likelihood of the model j and $P(\theta_j|M = j)$ is the prior distribution of the parameter under model j. Here, y is independent on $\theta_{i \neq j}$ given that $M = j$ $(j = 1, 2, ..., k)$. As mentioned previously, M is a model indicator and for the given M, various θ_j's are assumed to be completely independent.

By defining $\pi_i = P(M = j)$ such that $\sum_{j=1}^{k} \pi_j = 1$, the joint distribution of y and θ when $M = j$ is as below.

$$P(y, \theta, M = j) = f(y, \theta_j, M = j) \times \pi_j \times \{\prod_{i=1}^{k} P(\theta_i|M = j)\}. \tag{11.8}$$

The following equation shows the full conditional independence of each θ_j and M.

$$P(\theta_j|\theta_{i\neq j},M,y) \propto \begin{cases} f(y|\theta_j,M=j)P(\theta_j|M=j) & \text{for } M=j, \\ P(\theta_j|M\neq j) & \text{for } M\neq j, \end{cases} \tag{11.9}$$

where $P(\theta_j|M\neq j)$ is called *"Pseudoprior"*. When $M=j$, we generate the graph from the usual model of the full conditional distribution and when $M\neq j$, we generate the graph from the linking density.

Hence for the model M, we have

$$P(M=j|\theta,y) = \frac{f(y|\theta_j,M=j)\prod_{i=1}^{k} p(\theta_i|M=j)\pi_j}{\sum_{n=1}^{k} f(y|\theta_n,M=n)\prod_{i=1}^{k} p(\theta_i|M=n)\pi_n}.$$

In the usual condition, the algorithm produces samples from the correct joint posterior distribution. In particular, the ratio

$$\hat{P}(M=j|y) = \frac{\text{the number of } (M^{(g)}=j)}{\text{total number of } M^g}$$

is a simple estimate to compute the Bayes factor between any two models while g denotes the number of samples. Thus, $(M^{(g)}=j)$ means the jth model for the gth sample.

11.3.3 Gibbs sampling

By using the Bayes theorem, a complete model for a joint density while $j=1,2,...$ can be written as

$$p(y,\theta^j,k) = p(y,\theta^k,k)p(\theta^1|\theta^2)..p(\theta^{k-1}|\theta^k)p(\theta^{k+1}|\theta^k)p(\theta^{k+2}|\theta^{k+1})... \tag{11.10}$$

In this expression, if we denote π_k as the prior distribution for the unknown dimension of a parameter k and $\pi_k(\theta^k)$ as the prior distribution for $\theta^k|k$, we can represent the joint distribution of the state y with a model parameter θ under the k dimension via

$$p(y,\theta^k,k) = p(y|\theta^k,k)\pi_k(\theta^k)\pi_k. \tag{11.11}$$

Here, to move between dimensions, we have infinite choices causing the precise probabilities to not be found. To solve the underlying challenge, Walker (2009) [12] introduces an auxiliary variable u which helps us have finite choices to move between dimensions. On the other hand, the latent variable u has a distribution in which $u=k$ with a probability q and $u=k+1$ with a probability $1-q$.

Since u depends only on k and the complete model can be stated as

$$p(u,y,\theta^j,k) = p(u|k)p(y,\theta^j,k). \tag{11.12}$$

Thereby, the steps of the algorithm can be listed as below:

1 Sample $\theta^{(k)}$ from $\pi_k(\theta^k|y,k)$.
Sample $\theta^{(k+1)}$ from $p(\theta^{k+1}|\theta^k)$ and sample $\theta^{(k-1)}$ from $p(\theta^{k-1}|\theta^k)$.

2 Sample u from some kind of a binomial distribution in which $p(u = k+1) = q$ and $p(u = k) = 1-q$.

3 For the given k, sample j which will be the next k, from the distribution below:

$$\begin{array}{ll} j = k|u = k+1 & \propto (1-q)p(y,\theta^{k+1},k+1)p(\theta^k|\theta^{k+1}). \\ j = k+1|u = k+1 & \propto qp(y,\theta^k,k)p(\theta^{k+1}|\theta^k). \\ j = k|u = k & \propto (1-q)p(y,\theta^k,k)p(\theta^{k-1}|\theta^k). \\ j = k-1|u = k & \propto qp(y,\theta^{k-1},k-1)p(\theta^k|\theta^{k-1}). \end{array}$$

where the sampling strategy is simplified by the following equality.

$$P(\theta^k|\theta^{k+1}) \times \pi_{k+1}(\theta^{k+1}) = P(\theta^{k+1}|\theta^k) \times \pi_k(\theta^k) \tag{11.13}$$

that is valid under the Gaussian Copula graphical model (see the appendix for the proof). So, the simplified version of the third step of the algorithm can be shown as follows.

$$\begin{array}{ll} j = k|u = k+1 & \propto (1-q)p(y|\theta^{k+1},k+1)\pi(\theta^{(k+1)}). \\ j = k+1|u = k+1 & \propto qp(y|\theta^k,k)\pi(\theta^{(k)}). \\ j = k|u = k & \propto (1-q)p(y|\theta^k,k)\pi(\theta^{(k)}). \\ j = k-1|u = k & \propto qp(y|\theta^{k-1},k-1)\pi(\theta^{(k-1)}). \end{array}$$

11.3.4 Quadratic approximation for sparse inverse covariance estimation (QUIC)

This algorithm [6] is suggested to estimate the inverse of a sparse covariance matrix where the data are Gaussian. In this calculation, there is a penalty term in the general formula related to the sparsity of the associated graph. By increasing the underlying term, the precision matrix which shows the structure of the conditional dependence structure between nodes, becomes more sparse.
Hereby, in the algorithm, let Y be an $(n \times p)-$dimensional data matrix and the sample covariance matrix is denoted by

$$S = \frac{1}{n-1} \sum_{k=1}^{n} (y_k - \hat{\mu})(y_k - \hat{\mu})^T, \tag{11.14}$$

where $\hat{\mu} - \frac{1}{n}\sum_{k=1}^{n} y_k$.

Given the regularization penalty term $\lambda > 0$, the regularized log-determinate is defined as below:

$$\arg\min\{-log|Y| + tr(SY) + \lambda \sum_{i,j=1}^{p} |Y_{ij}|\}. \tag{11.15}$$

In this expression, $tr(.)$ shows the trace of the matrix and $|.|$ denotes the determinant as used previously. Then, an algorithm computes the optimal Λ by taking the $(p \times p)$-dimensional empirical covariance matrix S, which is positive semi-definite, and the regularization matrix Λ as inputs and initializing Y on the first iteration via $Y_0 > 0$ and repeat the steps until the convergence. More details can be found in the study of Hsieh et al., (2014) [6].

11.4 Copula

Copula is a very broad concept that is used to model the relationship between two or more variables. Generally, it means the connection that is used in statistics as a method to express the joint distribution between variables by increasing the dimension that is the number of variables. Copula was introduced first by Sklar (1959) [11] as a theorem presented below.

Let F be the d-dimensional distribution function of the random vector $Y = (Y_1, Y_2, .., Y_d)^T$ with margins $F_1, F_2, .., F_d$. Then, there exists a copula C such that for all $y = (y_1, y_2, .., y_d) \in (-\infty, +\infty)^d$, $F(y) = C(F_1(y_1), .., F_d(y_d)$. C is unique if $F_1, F_2, .., F_d$ are continuous.

To explain the way of using this theorem in application, it is enough to find the marginal cumulative distribution function (for each variable) and then use it as a uniformly distributed variable in the copula function to make a joint distribution function that is the multiplication of the marginal density function and the copula term.

There are more than two families of defined Copula in the literature. But the most famous ones in application are Archimedean copulas and Elliptical copulas.

11.4.1 The Elliptical copulas

This family contains only Gaussian and Student-t distributions which are symmetric with one and two parameters, in order. The table below represents some of the properties of Elliptical copulas for the bivariate model. We know from the multivariate joint distribution function of Gaussian and Student-t that the parameters are in the shape of vector or matrix.

The General vision of Elliptical copulas are in the form of $C(u_1, u_2, ..., u_d) = F(F_1^{-1}(u_1), F_2^{-1}(u_2), ..., F_2^{-1}(u_d))$ where u_is are from the uniform $(0, 1)$ distribution and copula term $C(u_1, u_2, ..., u_d)$ is d-dimensional Gaussian and Student-t distribution for this copula family. So, to use one of the Elliptical copulas, the in-

Table 11.1: The denotation and properties of the bivariate Elliptical families.

	Distribution	Parameter range	Kendall's τ	Tail dependence
1	Gaussian	$\rho \in (-1,1)$	$\frac{2}{\pi}\arcsin(\rho)$	0
2	Student-t	$\rho \in (-1,1), \nu > 2$	$\frac{2}{\pi}\arcsin(\rho)$	$2t_{\nu+1}(-\sqrt{\nu+1}\sqrt{\frac{1-\rho}{1+\rho}})$

verse of the cumulative distribution function should be found so that these values can be applied as the values of multivariate Gaussian and Student-t distribution.

It is seen from Table 11.1 that the only parameter of Gaussian copula is ρ which is the correlation coefficient between the variables that can be estimated not only by the maximum likelihood estimation (MLE) method, but, also by using its one-to-one relation with Kendall's τ. For example if the non-parametric correlation coefficient between two variables is $\tau = -0.5$, can be estimated as $\sin(-0.5 \times \frac{\pi}{2}) = -0.48$ regarding the relation of Kendall τ: $\frac{2}{\pi}\arcsin(\rho)$ as presented in Table 11.1.

In Student-t copula, apart from the correlation coefficient , there is the tail dependence as well. To explain the tail dependence that has been introduced as heavy tailed distribution in elementary level, let say when two variables Y_1 and Y_2 are tail-dependent. It means that in the very small or very large values of Y_1, there is some expectancy to face very small or very large values for Y_2 and vice versa. The tail dependent can be one or two sided, as well. One of the advantages of using the copula is to catch the tail dependence, which is not negligible, especially, in the sensitive data sets. The tail dependence can be seen clearly in Figure 11.1 which is in terms of the copula parameters, the correlation parameter ρ and the degrees of freedom ν. As can be observed in the figure, while ν increases, t-copula converges to a Gaussian copula and the tail dependence value goes to zero.

Furthermore, in Figure 11.1, we represent an example of a copula CDF and PDF of a bivariate student-t copula with a dependence parameter $\rho = 0.7$ and degree of freedom 4. Additionally, Figure 11.2 shows the PDF and scatter plot of a simulated Gaussian copula for different values of τ in order to emphasize the effect of the correlation in the elliptical copula family structure.

11.4.2 The Archimedean copula

As mentioned earlier, the second copula family are the archimedean copulas that are in the form of

$$C(u_1, u_2, ..., u_d) = \psi^{[-1]}(\psi_1(u_1) + \psi_2(u_2) + ... + \psi_1(u_d)), \qquad (11.16)$$

where ψ is the generator function and $\psi^{[-1]}$ refers to the inverse function of positive values. We present the generator function in Table 11.2. To clear the relationship between the generator function and the bivariate copula, we can

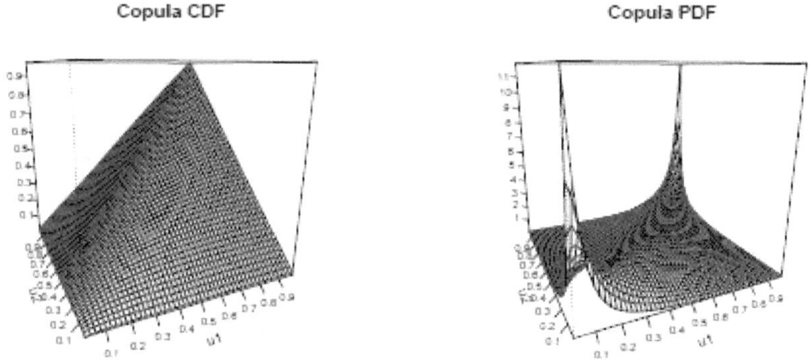

Figure 11.1: CDF and PDF of a bivariate student-t copula [1].

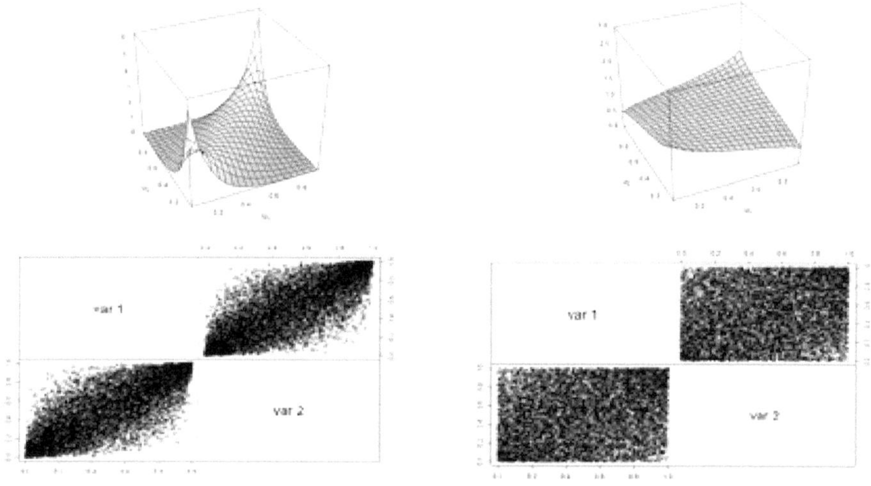

Figure 11.2: PDF and the scatter plot of the simulated Gaussian copula for $\rho = 0.8$ (left) and $\rho = 0.2$ (right).

give an example in the Clayton family. Let the generator function be taken as $\psi(t) = \frac{1}{\theta}(t^{\theta} - 1)$. So the inverse of the generator function can be written as $\psi^{-1}(t) = (t\theta + 1)^{-\frac{1}{\theta}}$ where t is replaced by $\psi(u_i)$'s to get the multivariate copula function. Herein, the pseudo-inverse is equal to the inverse when the inverse is not negative which is equivalent to the rectifier function. In order to get the copula formula, the bivariate Clayton applies uses Equation 11.16 and puts u_1 and u_2 in it, i.e., $C(u_1, u_2) = \psi^{[-1]}(\frac{1}{\theta}(u_1^{-\theta} - 1) + \frac{1}{\theta}(u_2^{-\theta} - 1))$ which is equal to $\max((u_1^{-\theta} + u_2^{-\theta} - 1), 0)^{-\frac{1}{\theta}}$. It can be done for more dimensions, as well. Therefore, we see that the generator function is simpler to write specifically for the archimedean families with more than one parameter such as BB1, BB6, BB7

Table 11.2: The notation and the properties of the bivariate archimedean families.

#	Name	Generator function	Parameter range	Tail dependence
3	Clayton	$\frac{1}{\theta}(t^{-\theta}-1)$	$\theta > 0$	$(2^{-\frac{1}{\theta}},0)$
4	Gumbel	$-(\log t)^{\theta}$	$\theta \geq 1$	$(0,2-2^{-\frac{1}{\theta}})$
5	Frank	$-\log(\frac{e^{-\theta t}-1}{e^{-\theta}-1})$	$\theta \in R$	(0,0)
6	Joe	$-\log(1-(1-t)^{\theta})$	$\theta > 1$	$(0,2-2^{-\frac{1}{\theta}})$
7	BB1	$(t^{-\theta}-1)^{\sigma}$	$\theta > 0, \sigma \geq 1$	$(2^{-\frac{1}{\theta\sigma}},2-2^{\frac{1}{\sigma}})$
8	BB6	$(-\log(1-(1-t)^{\theta}))^{\sigma}$	$\theta > 0, \sigma \geq 1$	$(0,2-2^{-\frac{1}{\theta\sigma}})$
9	BB7	$(1-(1-t)^{\theta})^{-\sigma}-1$	$\theta \geq 1, \sigma > 0$	$(2^{-\frac{1}{\sigma}},2-2^{\frac{1}{\theta}})$
10	BB8	$-\log(\frac{1-(1-\sigma\theta)^{\theta}}{1-(1-\sigma)^{\theta}})$	$\theta \geq 1, \sigma \in (0,1)$	(0,0)

and BB8, that stand for the Clayton-Gumbel, the Joe Gumbel, the Joe-Clayton and the Joe-Frank copulas, respectively. These families with more than one parameter (two-parameter families) are made by a combination of one parameter archimedean families to provide a more flexible structure like covering one and two-sided tail dependence and also most of them are appropriate for a non-symmetric joint distribution. Table 11.2 shows the generator function and also some other properties of the members of the archimedean copula families.

As an example of the archimedean copula family. Figure 11.3 represents some one-parameter bivariate archimedean copula families with the same Kendall's τ.

On the other hand, all of the copula families can rotate regarding the best fit for the data set according to the following equations. Because unlike Elliptical copulas, they are not necessarily symmetric.

$$\begin{aligned} C_{90}(u_1,u_2) &= u_2 - C(1-u_1,u_2), \\ C_{180}(u_1,u_2) &= u_1 + u_2 - 1 + C(1-u_1,1-u_2), \\ C_{270}(u_1,u_2) &= u_1 - C(u_1,1-u_2), \end{aligned}$$

where C_i represents the i degrees rotated version of the copula for $i = 90, 180, 270$. Figure 11.4 indicates an example of the rotation for the bivariate Clayton copula with Kendall's $\tau = 0.5$.

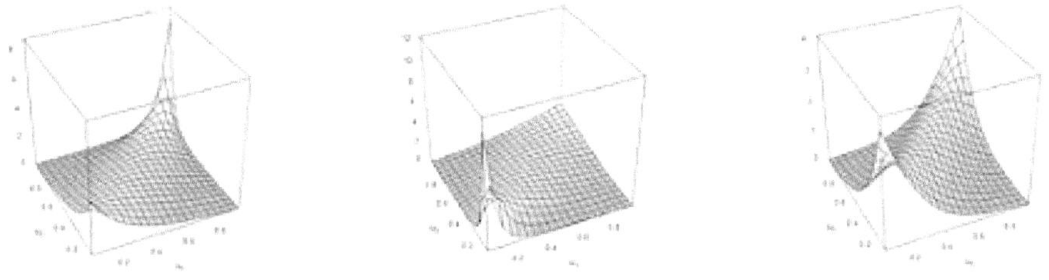

Figure 11.3: Gumbel, Clayton and Frank copulas, respectively, from left to right with parameters corresponding to Kendall's τ values of 0.5.

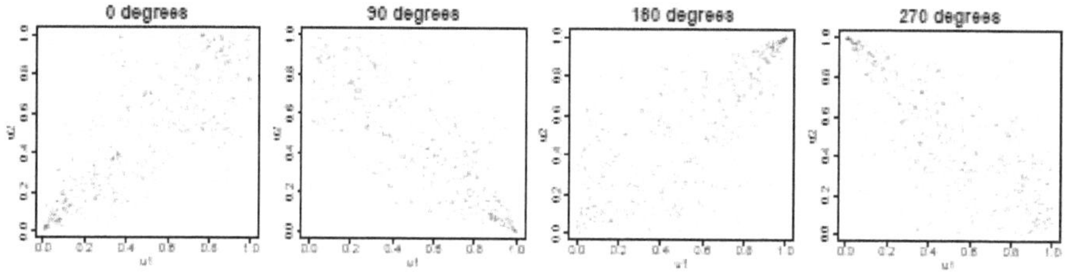

Figure 11.4: Samples from the Clayton copulas rotated with Kendall's $\tau = 0.5$ [1].

11.5 Vine Copula in Inference of Complex Data

As the model dimensionality increases, the model becomes more complicated and in most cases, the relationship between every two variable is not the same for all of the variables which means all of the edges cannot be modeled by a specific copula family. As a solution to decrease the complexity and model each pair freely, vine copula is proposed so that we can divide a d-dimensional joint distribution function into $\frac{d(d-1)}{2}$ bivariate functions. In order to illustrate the pair copula construction model, we give an example for $d = 3$ as below. The model dimension is three since there are three variables Y_1, Y_2, Y_3. Their joint distribution function is

$$f(y_1, y_2, y_3) = f_1(y_1) f(y_2|y_1) f(y_3|y_2, y_1).$$

$$f(y_2|y_1) = \frac{f(y_1, y_2)}{f_1(y_1)} = \frac{c_{1,2}(F(y_2)F(y_1)) f_1(y_1) f_2(y_2)}{f_1(y_1)} = c_{1,2}(F(y_2)F(y_1)) f_2(y_2). \tag{11.17}$$

Accordingly,

$$f(y_3|y_2, y_1) = c_{2,3|1}(F(y_2|y_1), F(y_3|y_1)) c_{1,3}(F_1(y_1), F_3(y_3)) f_3(y_3). \tag{11.18}$$

By combining Equation 11.19 and 11.17, we have:

$$\begin{aligned} f(y_1, y_2, y_3) = f_1(y_1) f_2(y_2) f_3(y_3) c_{1,2}(F(y_2)F(y_1)) \\ c_{1,3}(F_1(y_1), F_3(y_3)) c_{2,3|1}(F(y_2|y_1), F(y_3|y_1)). \end{aligned} \tag{11.19}$$

Therefore, the whole structure is shown in terms of pair copulas in a way that each of them can be investigated independently. It means in Equation 11.19, instead of working with three variables Y_1, Y_2 and Y_3 at the same time (shown as 1-2-3), we work with 1-2, 1-3 and 2-3—1 in one scenario, 1-2, 2-3 and 1-3—2 in another scenario and also changing the order of variables which can be written as a model itself. So, there are $\frac{p(p-1)}{2}$ different ways to write a multivariate structure in terms of pair copula when p denotes the number of variables, also known as dimension. Therefore, one of the most challenging issues becomes the selection of the best model among all available models.

In the case of analysis by the pair copula, we have three phases: a regular vine, a canonical vine and a drawable vine shown briefly as R, C, and D-vine, respectively. Here, each has a different structure. R-vine is a general form of vine copula which can be written as a combination of the other two models. Their structures are different. But both of them deal with pair-copulas. It depends on the way the joint distribution function is written and a good way to show their structure and difference is by their graphical representation in Figure 11.5 for illustration. This example is for a data set with 4 variables. In these graphs $X,Y|Z$ form denotes the structure of X and Y variables give the structure of Z. On the other side, X,Y form refers to the structure of X and Y variables together without any condition.

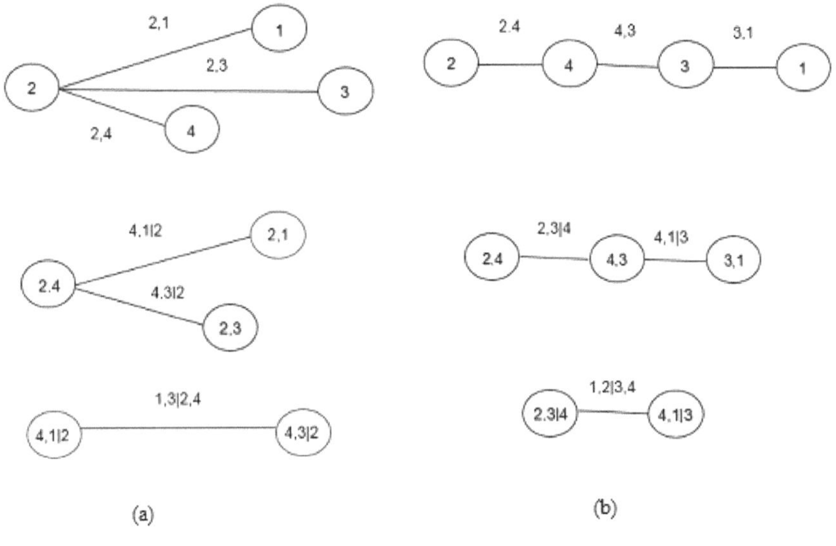

Figure 11.5: Examples of a four-dimensional C-Vine (a). left panel with order $\{2,4,1,3\}$) and a D-vine structure (b). The right panel with order $\{2,4,3,1\}$)).

In the C-vine, the first edge (variable) is selected in the first tree, the second root is selected in the second tree and so on. Hence, for d-dimensional data, we should choose $d-1$ roots for every tree. The order of variables in the left side of Figure 11.5 is $\{2,4,1,3\}$.

a. In the first figure, we see the possible connection of 2 with 1, 3 and 4. Then, among alternatives (2,4) structure is chosen and this structure is binded either to 1 (as 4,3—2). Between them, 4,1—2 is chosen and therefore, in the bottom figure, this structure is binded to 3.

b. In the first figure the line structure of connection is seen. Once 2,4 is binded, 3 and as seen in in the bottom Figure, 1 is binded to the structure.

In the D-vine copulas, the nodes of the first tree determine the whole model.

And finally, both vine copulas look like a vine and they include bivariate cases, only. The order of variables in the right side of Figure 11.5 is $\{2,4,3,1\}$ as example of C-vine copula.

In each ith tree $i = 1,2,...,d-1$, there are $d-i+1$ nodes (links) and $d-i$ edges. Each edge can be represented by an appropriate copula. The empirical copula which is used in data analysis with the copula approach is defined as

$$C_n(u) = \frac{1}{n}\sum_{i=1}^{n} 1(\hat{U}_i < u), u \in [0,1]^d$$

where $\hat{U}_i = R_i/(n+1)$ in which $R_i = (R_{i1},...,R_{id})$ and R_{ij} is the rank of Y_{ij} among $Y_{1j},...,Y_{nj}$ for $i = 1,2,...,n$ and $j = 1,2,...,d$. This transformation makes the data in the interval of $(0,1)$ to be used as a copula data. In Equation 11.19 which is represented as the simplest form of the joint distribution for a three cases function, it can be seen that the joint distribution could be decomposed to some pair copula terms which consist of some of the conditional term and the remaining not-conditional terms without any specific assumption or elimination.

There are several ways to construct the best model. As mentioned earlier, a regular vine is a general form of the vine copula that can be a C-vine in the tree and D-vine in another tree or even their combinations. So in every data set, there are $\frac{p(p-1)}{2}$ edges that should be determined with a pair copula from different families and from any rotation for each pair copula. There are some tests that determine the best-fitted pair copula for every two variables (given some others) as well as some goodness of fit tests. There are also some tests that compare two models based on the maximum-likelihood values [1]. The diagram in Figure 11.6 shows the process of the analysis by considering the available suitable functions in **R**. There are several functions in the "CD-vine" or its last version "VineCopula" package in **R** to be used in analyzing the data by vine copula. But the function which is written in the diagram, is the most important ones. In Table 11.3, their action is indicated very briefly.

On the other hand, the order of variables determines the root of each tree in the C-vine and the path in the first tree in the D-vine. The algorithm of the order selection is briefly described as follows:

- Compute the empirical distribution function of the data to transform them into uniformly distributed data.
- Compute the Kendall's τ correlation coefficient of the new data and select the variable with the largest τ as the first root.
- Select the best copula for each node between the first root and other variables and then, estimate the parameter(s). (The model parameters can be estimated by the MLE method for one or two-parameter families and by Kendall's τ in only one-parameter copula families.)

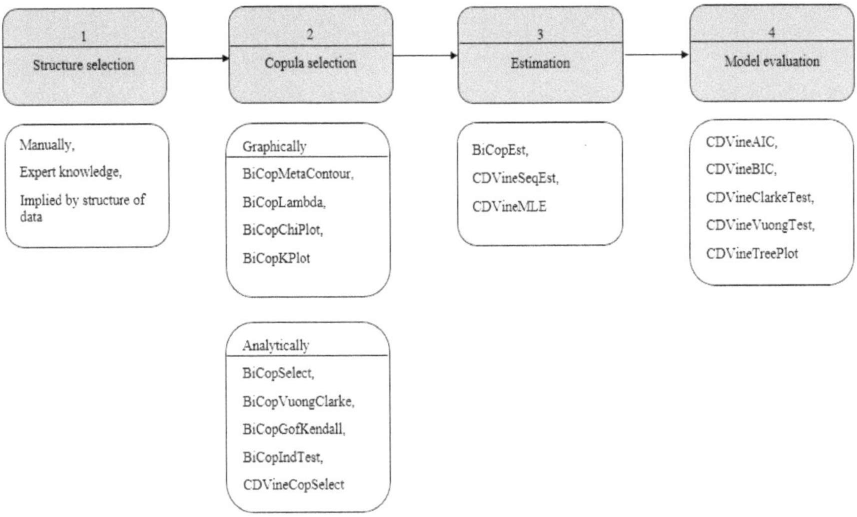

Figure 11.6: Proposed data analysis and model building work by some functions in package **VineCopula** in **R**.

Table 11.3: The most important functions' names and descriptions according to the VineCopula package in R programming language [1].

	Name	Explanation
1	BiCopMetaContour	Contour Plot of Bivariate Meta Distribution.
2	BiCopLambda	Lambda-Function (Plot) for Bivariate Copula Data.
3	BiCopChiPlot	Chi-plot for Bivariate Copula Data.
4	BiCopKPlot	Kendall's Plot for Bivariate Copula Data.
5	BiCopSelect	Selection and MLE of Bivariate Copula Families.
6	BiCopVuongClarke	Scoring GOF Test based on Vuong And Clarke Tests.
7	BiCopGofTest	Goodness-of-Fit Test for Bivariate Copulas.
8	BiCopIndTest	Independence Test for Bivariate Copula Data.
9	RVineClarkeTest	Clarke Test Comparing Two R-Vine Copula Models.
10	BiCopEst	Parameter Estimation for Bivariate Copula Data.
11	RVineSeqEst	Sequential Estimation of an R-Vine Copula Model.
12	RVineMLE	Maximum Likelihood Estimation of an R-Vine Copula Model.
13	RVineAIC	AIC and BIC of an R-Vine Copula Model.
14	RVineClarkeTest	Clarke Test Comparing Two R-Vine Copula Models.
15	RVineVuongTest	Vuong Test Comparing Two R-Vine Copula Models.
16	RVineTreePlot	Visualization of R-Vine Tree Structure.

- Transform the data by conditioning to the first selected variable via the h function by using the parameters estimated from the previous step as

$$h(x|\delta,\theta) := \frac{\partial C_{x\delta_j(F(x|\delta_{-j}),F(\delta_j|\delta_{-j})|\theta)}}{\partial F(\delta_j|\delta_{-j})}$$

  where δ is the estimated parameter(s) in the previous step and δ_j refers to an arbitrary component.

- Select the variable among the new data that has the largest Kendall's τ as the second root.

- Continue the process until the $(d-1)$th root is found.

The algorithm is somehow similar to the forward selection in the multiple regression. To estimate the model parameters by the maximum likelihood estimation (MLE), the order and the copula families should be determined before. There are some methods to select the best pair copula between the nodes such as graphical tools like the contour plot and some other statistical tests like the Voung and Clarke or the goodness of fit test. In Voung and Clarke, the test compares all available choices two by two and gets a score for each family if it was better than its alternative. The family with the maximum score is chosen as the best one and then, another goodness of fit test based on a χ^2 statistics determines if the proposed family is well fitted or not by giving a p-value. Similarly, there are some other tools to compare two models by using the AIC and BIC criteria or the Vungo test as well. All of the details for the mentioned tests are available in [1].

11.6 Application

To evaluate the performance of the RJMCMC alternatives, we compare the estimated networks of one real network. Here, the used dataset is the Rochdale data shown in Table 11.4 which is a binary data collected from 665 samples to assess the relationship among eight factors affecting women's economic activity. The eight variables are "a" to "h" and all of them are in the binary (yes/no) format. The cells appear row by row in a lexicographical order.

For instance, the first cell of Table 1.1 shows that 5 of the 665 persons, a=1, b=1, c=1, d=1, e=1, f=1, g=1, h=1 and also, for 57 persons, a=2, b=1, c=1, d=1, e=2, f=2, g=1, h=1.

In our analyses, we compare the F_1-score and Matthew's correlation coefficient (MCC) of RJMCMC, BDMCMC, QUIC and Gibbs Sampling with the true network given in the study of Whittaker (1997) [13]. The expression of the F_1-score and MCC are given as below.

$$F_1 - \text{score} = \frac{2\text{TP}}{(2\text{TP+FP+FN})},$$

Table 11.4: Rochdale data.

5	0	2	1	5	1	0	0	4	1	0	0	6	0	2	0
8	0	11	0	13	0	1	0	3	0	1	0	26	0	1	0
5	0	2	0	0	0	0	0	0	0	0	0	0	0	1	0
4	0	8	2	6	0	1	0	1	0	1	0	0	0	1	0
17	10	1	1	16	7	0	0	0	2	0	0	10	6	0	0
1	0	2	0	0	0	0	0	1	0	0	0	0	0	0	0
4	7	3	1	1	1	2	0	1	0	0	0	1	0	0	0
0	0	3	0	0	0	0	0	0	0	0	0	0	0	0	0
18	3	2	0	23	4	0	0	22	2	0	0	57	3	0	0
5	1	0	0	11	0	1	0	11	0	0	0	29	2	1	1
3	0	0	0	4	0	0	0	1	0	0	0	0	0	0	0
1	1	0	0	0	0	0	0	0	0	0	0	0	0	0	0
41	25	0	1	37	26	0	0	15	10	0	0	43	22	0	0
0	0	0	0	2	0	0	0	0	0	0	0	3	0	0	0
2	4	0	0	2	1	0	0	0	1	0	0	2	1	0	0
0	0	0	0	0	0	0	0	0	0	0	0	0	0	0	0

Table 11.5: The comparison between accuracy of different methods.

Methods	TP	FP	FN	TN	F_1-score	MCC
RJMCMC	14	1	0	14	0.96	0.866
BDMCMC	11	8	3	9	0.70	0.322
Gibbs	12	0	2	14	0.92	0.866
QUIC ($\lambda = 0.12$)	14	1	0	13	0.96	0.930

where TP is the numbers of truely found edges, FP presents the numbers of falsely found edges and FN denotes the numbers of edges that exist, but, are not recognized. Its perfection level is 1 and the range lies from 0 to 1.

$$\text{MCC} = \frac{(\text{TP} \times \text{TN}) - (\text{FP} \times \text{FN})}{\sqrt{(\text{TP+FP})(\text{TP+FN})(\text{TN+FP})(\text{TN+FN})}}.$$

Matthew's correlation coefficient is also known as the *phi coefficient* and turns a value between -1 and $+1$. A coefficient $+1$ represents a perfect prediction, 0 implies no better than random prediction and -1 indicates total disagreement between prediction and observation.

In our analyses, the number of iteration for RJMCMC, BDMCMC and Gibbs taken as 10^6 and for QUIC is only 1000 iterations. From the result, it is seen that RJMCMC and QUIC have the highest accuracy via F_1-measure, whereas, QUIC is the best under the MCC measure. On the other hand, the Gibbs algorithm is the second best in both F_1-measure and MCC in the construction of the undirected network for the Rochdale data.

On the other side, we implement R-vine approach in the inference of this data, too. To be used in copula methodology, we transformed the data to the Gaussian data through a method suggested by Hoff (2007) [5] and the R-vine method was applied to this latent data to see the relationship the variables named from a to h. According to our proposed method, R-vine, the significant non-zero edges are estimated as ef, dg, cg, cf, ce, bh, be, bd, ag, ae, ad, cd meaning that 10 of 13 relationships are estimated correctly and cd is the overestimated edge. By these outputs, we computed TP=10, FP=1, FN=3 and TN=14. Accordingly, F_1-score= 0.83 and MCC=0.72 while in the work done by Purutçuoğlu and Farnoudkia (2017) [10], these values are found as F_1-score= 0.96 and MCC=0.86, respectively.

11.7 Discussion

Based on the represented accuracy measures for the different methods, it is seen that among Bayesian methods, RJMCMC performs better than BDMCMC and Gibbs sampling though it needs more computational time. Whereas, QUIC outperforms not only in accuracy, but also, in computational time. However, QUIC is somehow, a non-parametric method that optimizes the target matrix based on some penalty term. Therefore, as the penalty term increases, the precision matrix becomes more sparse. So, to obtain a specific sparsity rate, we need to determine the penalty term from the beginning which needs some background that is not always available. The other Bayesian methods also needs some threshold to estimate the adjacency 0 and 1 matrix that is easier to decide. In the second part of this chapter, the Copula method was discussed in terms of its advantage and challenges and we observed its high accuracy in comparison with RJMCMC alternatives. Besides the high performance of the R-vine copula, it can catch hidden facts between every two variables unlike the other methods discussed in this chapter. As further work, we consider the highly correlated data and also some likelihood-based methods that need more samples in comparison with the proposed methods for this study. It can be also, the directed network in which the relationship between two variables is not always two-sided, i.e., one sided, too. These kinds of networks are important, particularly when the time order matters.

References

[1] E. Brechmann and Ulf. Schepsmeier. Cdvine: Modeling dependence with c-and d-vine copulas in R. *Journal of Statistical Software*, 52(3): 1–27, 2013.

[2] B.P. Carlin and S. Chib. Bayesian model choice via markov chain monte carlo methods. *Journal of the Royal Statistical Society: Series B (Methodological)*, 57(3): 473–484, 1995.

[3] A. Dobra and A. Lenkoski. Copula gaussian graphical models and their application to modeling functional disability data. *The Annals of Applied Statistics*, 5(2A): 969–993, 2011.

[4] P.J. Green. Reversible jump markov chain monte carlo computation and bayesian model determination. *Biometrika*, 82(4): 711–732, 1995.

[5] P.D. Hoff. Extending the rank likelihood for semiparametric copula estimation. *The Annals of Applied Statistics*, 1(1): 265–283, 2007.

[6] C.-J. Hsieh, M.A. Sustik, I.S. Dhillon and P. Ravikumar. Quic: Quadratic approximation for sparse inverse covariance estimation. *The Journal of Machine Learning Research*, 15(1): 2911–2947, 2014.

[7] V. Mazet and D. Brie. An alternative to the rjmcmc algorithm. In *2006 IAR Annual Meeting, Nancy, France*, pp. CDROM. IAR, 2006.

[8] A. Mohammadi and E.C. Wit. Bayesian structure learning in sparse gaussian graphical models. *Bayesian Analysis*, 10(1): 109–138, 2015.

[9] B. Muthén. A general structural equation model with dichotomous, ordered categorical, and continuous latent variable indicators. *Psychometrika*, 49(1): 115–132, 1984.

[10] V. Purutçuoğlu and H. Farnoudkia. Gibbs sampling in inference of copula gaussian graphical model adapted to biological networks. *Acta Physica Polonica, A.*, 132(3): 1112–1117, 2017.

[11] M. Sklar. Fonctions de repartition an dimensions et leurs marges. *Publication Institute Statistics University of Paris*, 8: 229–231, 1959.

[12] S.G. Walker. A gibbs sampling alternative to reversible jump mcmc. *arXiv preprint arXiv:0902.4117*, 2009.

[13] J. Whittaker. *Graphical Models in Applied Multivariate Statistics*. Wiley Publishing, 2009.

Chapter 12

Evaluation of Data Compression Methods for Efficient Transport and Classification of Facial EMG Signals

Fikret Arı,[1] *Erhan Akan,*[2,*] *Hayriye Aktaş Dinçer,*[3] *Ekin Can Erkuş,*[4] *Mahdieh Farzin Asanjan,*[4] *Didem Gökçay,*[5] *Fatih Ileri,*[2] *Vilda Purutçuoğlu*[4] and *Abdullah Nuri Somuncuoğlu*[4]

12.1 Introduction

In recent years, due to advancements in technology and processing power, the utilization of digital biomedical data became favorable for medical diagnosis and treatment. This increment demands improvement of the data processing and transportation techniques. One of the main issues which arise is the size of the recorded data which is proportional to the data resolution and recording duration. For example, a usual wired EEG setup is generally composed of 64 sensors and a

[1] Ankara University.
[2] Turkish Aerospace Industries.
[3] Fatih Sultan Mehmet Vakıf University.
[4] Middle East Technical University.
[5] 21yy Ltd.
All authors have equal contribution to this study.
* Corresponding author: erhanakan@gmail.com

very high sampling frequency ($\sim$ 2.5 kHz), also, ECG comprises 4 sensors with a lower sampling frequency at the range of a few hundred of Hertz [4]. Hence, the compression of the data to decrease the size of the record without losing any important portion of the recorded data seems to be a crucial optimization problem in biomedical research. Another aspect that needs consideration is the missing parts (records) in the data because in some cases it makes data interpretation impossible. Therefore, before any processing, the data imputation/interpolation in an optimal way seems to be unavoidable. Finally, data classifications seem to play an important part in the optimal procedure of data processing because recent artificial intelligence and machine learning approaches call for the automatic classification of data to aid diagnosis or treatment. In this chapter, our focus will be on the compression of EMG signals without losing accuracy in data classification.

Since the EMG data is recorded over a long period of time, the storage or the transportation of these signals require compression for efficiency. There are two types of data compression techniques comprising lossless and lossy methods. In these two approaches, the former can compress and decompress the data to obtain the exact original data. The latter, however, can include a slight amount of data to be lost during the decompression, allowing for an approximation of the original data to be retrieved. Especially, when the lost parts cover the noise or the unwanted frequency within the signal record, the approximation of the original data favors the relevant parts of the data stream. In general, compression of signals can be performed either in the time domain or transformed domain [22]. There are several optimization methods for data compression which have their own advantages and disadvantages such as the pattern matching method, which is based on the detection of the repetitive pattern of the data and uses the pattern to apply the compression [15], and the utilization of the Wavelet transformation [27, 14, 2].

Finally, classification of EMG signals can be achieved by using several techniques of pattern recognition within the biomedical records. The most famous methods are decision tree, naïve Bayes, neural network, bagging, kernel density, and support vector machines algorithms [46]. Additionally, K-Nearest Neighbor (K-NN) is a well-known non-parametric classification method in operational research, successfully applied for data mining of the biomedical data (i.e., [49]). On the other side, fully expanded decision tree classifiers and Pruned decision tree classifiers are 133 errors803 warnings successfully implemented to both cost reduction of the diagnosis and improvement in the efficiency of this process [40] as an optimal solution of the underlying challenge.

In the following section, we will provide background information on the characteristics of the EMG signal, as well as information about a few data compression methods, data classification techniques, and applications relevant to EMG. Then, we will illustrate a use case in which two EMG compression methods, the Discrete Cosine Transform [28] and PCA [41], two common data compression methods in operational research, are implemented to achieve the minimal loss in classification accuracy while saving space and time during signal transportation and storage. Finally, we will discuss very briefly which data compression method will be favorable under which circumstances.

At this point, it is important to emphasize that the compression we present will focus on the performance of classification of the EMG signals, not a complete reconstruction of the original EMG signal with the smallest error rate.

12.2 Background

12.2.1 Characteristics of the EMG signal

Electromyography (EMG) is a technique to observe the electrical potential of muscle activity. Contraction of the muscles causes electrical potentials which can be measured by EMG [35]. The frequency of a typical EMG signal varies from a few Hz to 500 Hz. Generally, a sampling frequency of 1 kHz [39] or 2 KHz [21] is recommended. The EMG signal frequency may overlap with the power line supplying the apparatus, so 50–60 Hz band in an EMG signal should be treated as noise and eliminated. The amplitude of an EMG signal is low (around several hundred millivolts), and it is required to amplify the signal minimally at the level of 50–100 K.

In some EMG applications, a signal envelope is more important. In addition, microcontrollers accept rectified positive signed voltage inputs. Therefore, some sensors produce rectified and enveloped signal output.

In the Figure 12.1, a sample signal output from a surface EMG sensor is shown.

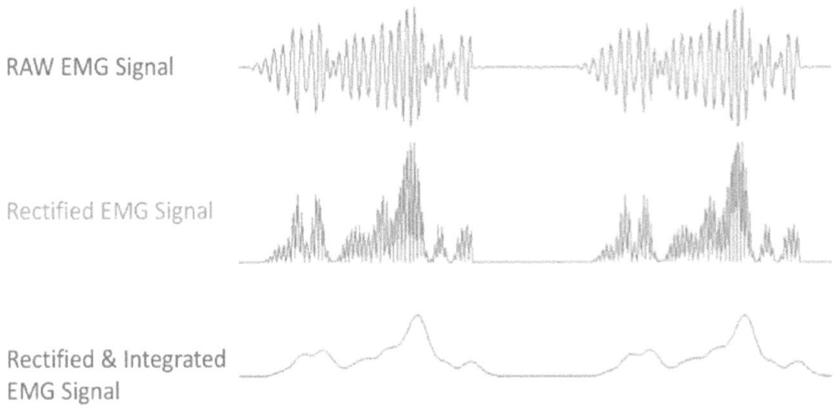

Figure 12.1: Signal output of an EMG device.

12.2.2 Compression and classification techniques used in EMG

Numerous techniques are used for EMG signal compression in the literature including wavelet transform [25, 8, 26], autoregressive (AR) modeling [12], image compression techniques [11, 17], and discrete cosine transform (DCT) [37].

The Fourier transform-based methods such as DCT were widely used for compression applications in the literature. Since DCT is real and the computation algorithm of DCT is fast [39], it is one of the most common techniques used for EMG data compression.

There are numerous studies in the literature dealing with the use of EMG signals in effective state classification and categorization. In a study published in 2013 dealing with EMG collected from facial muscles, a neural network architecture was developed in a new functionality setup [23]. In the related study, the aforementioned artificial neural network was trained by extracting 10 features from the EMG measurements.

In another study, in which state estimation was made from EMG measurements, it was aimed to determine stress from EMG signals collected from facial muscles of participants [36]. In the study, 5 coefficient vectors were obtained by applying 4-stage discrete wavelet transform (DWT) decomposition through the low and high pass filters of DB-7. The mean and standard deviation of these coefficient vectors extracted from each 300 millisecond EMG signal segment were used as feature vectors of these signal segments.

12.2.3 Applications of EMG

EMG is used for various areas including gait studies and rehabilitation [34], control of prosthetic devices [32], detection and monitoring of neuromuscular diseases [18], prevention of premature birth [16], preoperative measurements [38], virtual reality gaming [43], and ergonomics [33]. Apart from these, one of the most frequent application areas of EMG is emotion recognition, which correlates with the affective state [24, 42, 44].

Emotions are represented by facial expressions related to specific combinations of facial muscle activity. Facial electromyography (fEMG) provides an objective measurement of facial muscle activity related to emotions. The electrical activity of the underlying facial muscle can be measured and interpreted via fEMG. Moreover, fEMG is advantageous to other alternative approaches since the muscle activities that cannot be visually perceived by an observer can be detected via fEMG. The widely studied muscles in fEMG are zygomaticus major, corrugator supercilli, lateral frontalis, levator labii superioris medial frontalis, orbicularis oculi, and masseter. Generally, the classification and the interpretation of the EMG signals are based on the calculation and thresholding of amplitude-based features (e.g., mean absolute value).

Research focusing on emotion revealed that activity in the corrugator supercillimuscle (located over the eyebrow and activated when furrowing the brow) is related to negative or unpleasant emotional valence [10]. On the other hand, the positive emotional experience was indicated by an increase in the activity of the zygomaticus major muscle (located in the cheek and activated during smiling) and orbicularis oculi (located beneath the outer corner of the eyes, also activated in smiling). Orbicularis oculi muscle also represents the startle eye blink response which is indicative in case of fear [24].

12.2.4 Optimization of cost and performance

In terms of performance, compression ratio (CR) is a widely used measure in the literature to indicate the benefits gained in storage and transportation. CR is calculated as the size of the original signal divided by the size of the compressed signal. Size is calculated by either the length of the time series or the storage on disk/RAM. The other performance measure that interests us in this study is classification accuracy. Generally, accuracy is defined as the total number of correctly classified datasets divided by the total number of datasets tested during classification.

In terms of cost, the overhead brought by the compression of the EMG signal is introduced during the compression step at the sender and the reconstruction (or decompression) step at the receiver. This is calculated as the computation time spent during run-time. Usually, specific low-level subroutines are devised to measure the computation time during compression and decompression.

12.3 Methods

12.3.1 EMG compression techniques to be implemented

12.3.1.1 Discrete cosine transform

The discrete cosine transform is widely used in EMG compression. DCT is a good technique to determine the effect of different frequency components on the given signal. The logic behind the compression is to eliminate components that represent a small percentage of the entire signal. Then, by taking the inverse DCT of the eliminated sequence, the original signal could be reconstructed by a small error. So the signal is compressed with a lossy approach.

Briefly stating, a discrete time EMG signal $x[n]$ defined in $n = 0, ..., N-1$ can be represented by only k coefficients represented by $y[k]$ with the formula in 12.1, where practically, k can be any number between 3 and 10.

$$y[k] = C(k) \sum_{n=0}^{N-1} (x[n] cos[\frac{\pi(2n+1)k}{2N}]) \tag{12.1}$$

In this formula, $c(0) = \sqrt{1/n}$ and $C(k) = \sqrt{2/N}$ for $k = 1 \ldots N-1$.

It is important to note that the EMG signal depicted by $x[n]$ here is obtained from a single muscle of an individual person. Hence, such EMG signals can be easily compressed on the fly, during a real-time application.

The compression is carried out by sorting the coefficients in descending order using the "sort" function of MATLAB®. This function will give the indices of the sorted coefficients. Then a threshold is chosen to depict the percentage of the original signal to be reconstructed during decompression. This threshold is computed based on the absolute values of the DCT coefficients. The number of DCT coefficients to be saved is found by dividing the sum of the absolute value of the selected coefficients by the sum of the entire set of DCT coefficients until the chosen threshold is reached. When this ratio reaches the desired percentile, the remaining coefficients are eliminated. Since the dominant frequency components of an EMG signal are the lower frequencies, after a certain index of the DCT matrix of the signal, it is possible to eliminate the remaining indices. We used the embedded "dct()" function of MATLAB to compress the EMG and the "idct()" function of MATLAB to reconstruct the compressed EMG signal. It is important to note that this is a lossy compression, hence, we admit from the beginning that only a specific percent of the signal will be secured. Hopefully, the reconstructed part of the signal is the part that aids the classification process. Because we do not want to lose the parts of the signal relevant for classification.

After finding the DCT coefficients, the compression ratio, CR, is calculated by coding the compressed EMG signal. The length of the compressed signal is indexed by a specific time sample, j, after which the samples are coded as zeroes. The length of the compressed signal would be j+1. Because an extra byte is used to report the index itself. Hence, the compression ratio is reported as $(N-1)/j+1$, where the original length of the EMG signal is $N-1$.

12.3.1.2 Principle component analysis (PCA)

PCA is a good technique to reduce a higher dimensional signal into a lower dimensional signal set by joining related components together and eliminating esoteric dimensions. For this purpose, a covariance matrix is generated from the EMG signal collected from multiple subjects, and eigenvalues are computed. Then, eigenvectors are generated from a selected set of bigger eigenvalues that are sorted from the largest to smallest. The signal is compressed in a lossy manner, after eliminating the eigenvectors that correspond to smaller eigenvalues. It has to be decided from the beginning, how much percentage of the variance in the signal will be kept after elimination of the smaller eigenvalues. Based on this value, the minimal number of eigenvalues are chosen, such that the eigenvectors that correspond to these eigenvalues will represent the depicted percent variance in the signal after reconstruction. In MATLAB environment, using "pca()" function, it is possible to obtain the eigenvectors in descending order according to their importance, their eigenvalues, reconstruction potential of eigenvectors, score matrix, and mean value of data. Eigenvalues represent the importance of each eigenvector.

To compute the compression ratio, CR, for the EMG signal, we need to divide the size of the original data matrix before PCA to the size of the data matrix after PCA. Since the size of the time series does not change after PCA, the compression ratio is dependent upon how many subjects were present and how many eigenvalues are chosen.

It is important to note that the EMG signals here are obtained from a single muscle of multiple subjects. Hence, reducing the original EMG time series matrix has to be done offline by pooling data from multiple subjects together. This does not allow for a real-time application. Furthermore, the original EMG signals adhere to a specific subject group. If this subject group must be changed, performing a new PCA compression is imperative.

12.3.2 Emotion classification techniques to be implemented

The feature extraction phase is a preamble to proceed into the classification phase. Features represent the data in unique aspects and properties, and, particularly, the statistical or spectral features constitute the most commonly used feature families for the analyses of the time series data [20]. While the statistical features tend to represent the time domain characteristics, the spectral features represent the frequency domain properties of the data [5]. Since EMG data have both time and frequency domain characteristics, some most commonly used features in both feature families are used to compare the discriminability performances of the DCT coefficients [19].

In this study, some statistical features include the most commonly used features in time series analyses: the statistical moments such as mean, variance, skewness, and kurtosis; and other components such as range, root mean square (RMS), and the mean of the absolute value. On the other hand, the frequency domain features are selected as band power, mean and median of the frequency domain, and bandwidth power, all up to the Nyquist frequency limit. Other features include DCT coefficients of the compressed data and features extracted from the reconstructed PCA time series.

The classification process includes the training and testing (i.e., validation) phases after the feature extraction phase is completed [7]. Among many machine learning (ML) classifiers, tree, and k-nearest neighbour (K-NN) classifiers are selected based on their performances in preliminary analyses for a small set of experimental conditions of the same data. Both classifiers aim to train the data to generate a classification model based on the features extracted from the data. Hence, the new data intervals or samples are assigned to the predicted group in the validation step according to the generated model [3]. In this study, the groups of the data intervals are already known; thus, the supervised machine learning approach is used for testing performances. Classification accuracy is computed by calculating the number of True Positives plus the number of True Negatives divided by all datasets used for testing.

12.3.2.1 Tree classifier

Tree classifier is one of the simplest machine learning classifiers suitable for binary classification with fast computational speed [45]. The tree classifier computation starts by setting the maximum number of splits parameter, which refers to the number of times the tree branches split starting from the root of the tree [45]. The computational complexity increases by increasing the number of splits while reducing the classification's error rate [9]. Then, the features are assigned to each split by a dividing value to classify the samples for that feature. That dividing value is selected such that the classification accuracy of the respective feature is maximized unless it is limited. This process is iteratively repeated for each feature to form a tree classifier model where each leaf of a branch refers to one group that the sample to be classified [45]. MATLAB offers a built-in function "fitctree" in its "Statistics and Machine Learning Toolbox(TM)" [48] for this purpose. Our empirical trials indicated that the maximum number of the split parameter can be set to 7 to use in the EMG data.

12.3.2.2 K-nearest neighbour (K-NN) classifier

K-NN algorithm is also one of the most commonly used supervised ML classifiers. The algorithm defines a K value, preferably an odd value, to decide the predicted label. The training set is selected, and the features are distributed into F dimensional coordinates, where F refers to the number of features used for each data segment [31]. Then, without building a model, the test samples are placed into the respective coordinates regarding the values of the features. The closest K samples are marked based on a distance formula (i.e., Euclidean). The label of the majority of the K samples is assigned to the test sample [6]. In this study, the computations are performed using the "fitcknn" function of the "Statistics and Machine Learning Toolbox (TM)" in MATLAB [48] and K is set to 5 empirically.

It is important to shuffle the given EMG data in a balanced way into training and test subsets. For this purpose, the validation can be done exhaustively using the Leave-one-out Cross Validation method. This is one of the simplest algorithms, where all data samples except one are put into the training group and the one data that is set aside is used for testing. This is repeated by leaving out each of the data segments, one at a time, and training for the rest of the data [13]. The leave-one-out algorithm is suggested to be used for small sets of data. We preferred this method because the dataset which we chose to use in this study was small.

12.3.3 Use case: Prediction of fear from EMG

In response to stress or fear, the sympathetic nervous system prepares the body for the fight-or-flight response. Therefore, energy is sent to the muscles. Under stress, the electrical activity of the muscles is much higher than in the resting state, so this can be picked up by the EMG signals. However, not all muscles are affected equally. The most preferred muscles for stress measurement are the trapezius muscle which is between the neck and shoulder and the facial muscles [47]. Orbicularis oculi is also a muscle used for this purpose in the facial muscles as shown in Figure 12.2.

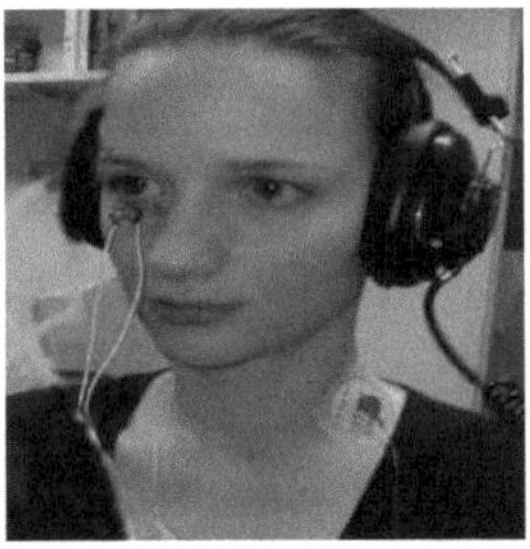

Figure 12.2: Positioning electrodes on the Orbicularis Oculi Muscle [1].

The main aim of this study is to observe how EMG compression affects EMG classification performance. For this purpose, we performed DCT and PCA compression on the EMG signals that were obtained in an emotion experiment and investigated the accuracies in classification. In this study, we used the EMG subset of a dataset from the Zenodo database (https://zenodo.org/record/3430920). The EMG data are part of a larger collection of physiological measurements, namely, PsPM-SMD: SCR, EMG, ECG, and respiration in response to auditory startle probes (Abbreviated as: Skin conductance response (SCR), orbicularis oculi electromyogram (EMG), electrocardiogram (ECG)). There are 19 participants (6 males and 13 females aged 24.9 +/- 4.1 years) who responded to 25 impulsive sounds [30].

In our study, we excluded the data of participants 8 and 12. Because the amplitudes of the impulse signal given to them were quite low and incomprehensible compared to the other participants. Furthermore, the first and last trials were eliminated since they did not contain meaningful data. So, only 23 trials from each participant were used. The intervals between the trials were randomly determined between 7–11 seconds for each participant as indicated in the supplement to the study "Modeling startle eyeblink electromyogram to assess fear learning" [29]). When we analyzed the dataset in more detail, we noticed that the impulse markers in the sixth structure in the dataset were incorrectly marked. For this reason, we corrected the sound markers based on the peak positions of the impulses in the impulse sound signals provided inside the downloaded Zenodo dataset.

In the experiment, it is expected that each impulsive sound would create a fearful eye muscle response in the participants. Obviously, the fearful response happens after the sound is inflicted, not before. Therefore, the entire EMG data stream can be split into fearful and non-fearful portions by extracting time series immediately after the impulse and immediately before the response. For this purpose, in the EMG datasets, 500 ms before the impulse signal and 500 ms after the impulse signal are used as the class inputs for the classification. The data before the impulses were classified as fear=0 (also referred to as resting), the data after the impulses were classified as fear=1.

Before the compression process, EMG data of 17 participants were segmented using the marker information in the Zenodo Dataset as mentioned above. 23 fearful stimulus responses and 23 non-fearful (or resting) EMG data segments were obtained from all participants. Due to the concerns related to noise and artifact, each of these events was cropped to contain only 500 milliseconds although EMG data were recorded for 5000 ms for each trial. After the data are prepared, we applied a low-pass filter with a cutoff frequency of 200Hz to the resting cases and stimulus responses one by one to get rid of the noise. In MATLAB environment, this was done using the "lowpass" function.

In the MATLAB environment, the absolute value of both the resting case and the event after the impulse is treated as separate signals to be compressed during the DCT compression. For the PCA, the same time series was used but without taking their absolute values. The cropped low pass filtered signals were concatenated into a single time series for each participant. This was done separately for the stimulus responses and resting cases. Thus, the data to be used in the "pca()" function was two matrices of size "total time duration of concatenated events" x "number of participants" (i.e., size = 11500x17).

In Figure 12.3, the reconstruction of a sample trial is shown. The top row contains the low pass filtered original EMG time series from a stimulus response (left) and resting (right) case that are labeled as fear and no-fear for classification. Rows 2 and 3 contain the reconstructed signals after compression with PCA or DCT, where graphs on the left correspond to the stimulus response (fear) and the graphs on the right correspond to the resting case (no fear). For compression with PCA, 6 eigenvectors were used. For compression with DCT, the reconstruction index threshold was set to 95 percent. As seen from the figures in rows 2 and 3, the reconstructions are quite successful.

Furthermore, we are able to observe that the stimulus response has a specific signature, while the resting case time series is non-specific. Due to this difference, for the same number of eigenvalues, reconstruction rates of the stimulus responses and resting cases will not be the same. This is because the reconstruction rates are based on variance and the stimulus responses have a specific signature that better represents the variance. In PCA, when 6 eigenvalues are used,

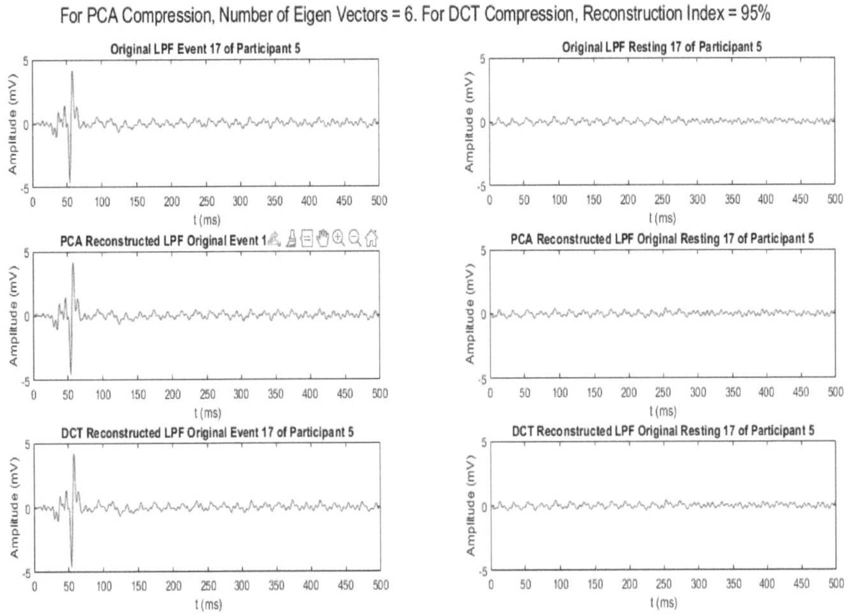

Figure 12.3: Reconstructed EMG signals after PCA and DCT compression.

the reconstruction rate is 92% for stimulus responses, but, it is approximately 85% for resting cases. For a similar reconstruction success in resting cases, more eigenvalues are needed for these time series.

A similar reconstruction success is observed in DCT as well. Due to the non-specific nature of the resting cases, the DCT coefficients that are higher than zeroes in the resting cases have much less magnitude compared to their counterparts in the stimulus response time series. After setting the reconstruction index threshold to a specific value, more DCT coefficients are needed to reach the reconstruction threshold to represent the resting case EMG signals compared to the stimulus responses. For instance, for the 95% reconstruction index threshold used in this figure, the number of DCT coefficients used to represent the resting cases are 1% more in comparison to those used for stimulus responses.

After the reconstruction of the EMG signals from their counterparts compressed with DCT or PCA, the same classification pipeline was executed to test the supervised classification performances. As discussed earlier, we used two classifiers: Tree versus k-NN. We tested 3 different feature sets: 1) features based solely on statistical/frequency characteristics of the EMG data, 2) features based solely on the coefficients of the Direct Cosine Transform (with 5 or 10 coefficients), or a combination of both.

12.4 Results

We will discuss the results of the binary classification process of predicting fearful versus non-fearful facial EMG recordings in terms of performance and computational cost separately.

12.4.1 Performance

The results we obtained from the original data are presented in terms of percent accuracy in Table 12.1. When the original EMG data are classified, the accuracy is between 81.9–84.4% on statistical features, 66.6–79.4% on DCT features and 84.0–81.5% on both feature sets. These scores improve slightly when the data are passed through a Low Pass Filter (LPF) of 200 Hz to get rid of noise. In that case, the performance of the statistical features varies between 84.0–85.0%, the performance of the DCT features vary between 76.3–81.0%, and the performance of combined features vary between 84.2–85.8%.

While compressing the EMG signals with DCT, we tried different levels of compression by setting the reconstruction index threshold to 95, 75, 50 and 25. After reconstruction, these lossy compressions resulted in CR values of 2.2, 2.4, 2.8, and 3.4 as seen in Table 12.2. For each of these reconstructed signals, 35, 22, 10 and 5% of the DCT coefficients are used to represent the stimulus responses, while slightly more number of coefficients are used to represent the resting cases. The accuracy of classification varied between 80.6% to 84.1% for statistical/frequency features and 82.2–86.0% for combined features including statistical/frequency and DCT coefficients. We observed a graceful degradation in the accuracy in classification with statistical/frequency features as CR increased, but, the accuracy did not change much with increasing CR when DCT coefficients are used as features exclusively. Overall, when all of the statistical, frequency, and DCT features are combined, reducing the signal energy from 95 to 25% improves the classification accuracy. While we save space through compression, we also obtain better classification accuracy. However, by allowing reconstruction of the signal at a 25% level, the EMG signal would become useless to utilize in applications other than classification.

While compressing the EMG signals with PCA, we tried different levels of compression by changing the number of eigenvalues from 1 to 17. To have results that we could compare with the DCT compression, we picked the eigen-

Table 12.1: The classification accuracy of ML classifiers for original or filtered data.

Data Type	Features used // Classifier (% Accuracy)							
	Stat + Freq		DCT 5		DCT 10		Stat + Freq + DCT 10	
	Tree	K-NN	Tree	K-NN	Tree	K-NN	Tree	K-NN
Original Data	84.40	81.97	66.62	71.74	74.30	79.41	84.02	81.59
200Hz LPF Data	84.01	85.02	76.34	79.52	77.94	81.05	84.27	85.81

Table 12.2: The classification accuracy of ML classifiers for DCT reconstructed data. (Compression and reconstruction are shortened to Cmprs and Rcnstr, respectively.)

Cmprs ratio	Rcnstr %	Remaining co-eff.	Features used // Classifier (% Accuracy)							
			Stat + Freq		DCT 5		DCT 10		Stat + Freq + DCT 10	
			Tree	K-NN	Tree	K-NN	Tree	K-NN	Tree	K-NN
2.2	95	35	83.63	84.14	75.57	77.11	76.98	80.69	83.76	83.24
2.4	75	20	83.25	83.82	76.21	79.16	77.24	80.33	83.25	84.02
2.8	50	10	81.05	82.74	75.44	78.01	79.41	81.84	85.81	83.89
3.4	25	4	80.69	82.61	75.96	78.26	79.28	80.82	86.06	82.23

values that corresponded to similar CR values. With 5, 6, 7, and 8 eigenvalues, we obtained CR values of 2.1, 2.4, 2.8, and 3.4 as seen in Table 12.3. For each of these reconstructed signals, the reconstruction percentages corresponding to the amount of variance kept after the lossy compression was 95, 94, 92, and 90% for the stimulus responses, while much lower reconstruction percentages were observed for the resting cases (more specifically, 93–80%). The accuracy of classification varied between 86.1% to 91.5% for statistical/frequency features and 86.8–91.3% for combined features including statistical/frequency and DCT coefficients. We observed a graceful degradation in the accuracy in classification with statistical/frequency features and also in classification with combined features as CR increased. Overall, the accuracy of classifications of EMG signals reconstructed from PCA compressions was higher in comparison to those reconstructed from DCT compressions. Furthermore, the reconstructed signals in PCA were much less lossy when compared to similar CRs of DCT compression.

Table 12.3: The classification accuracy of ML classifiers for PCA reconstructed data. (Compression and reconstruction are shortened to Cmprs and Rcnstr, respectively.)

Cmprs Ratio	Rcnstr %	#eigen values	Features used // Classifier (% Accuracy)							
			Stat + Freq		DCT 5		DCT 10		Stat + Freq + DCT 10	
			Tree	K-NN	Tree	K-NN	Tree	K-NN	Tree	K-NN
2.1	95	8	91.56	89.13	73.38	80.69	80.43	85.04	91.43	88.75
2.4	94	7	88.49	87.60	74.17	79.54	81.20	84.91	89.64	87.21
2.8	92	6	86.97	87.21	72.38	77.24	78.26	80.69	87.34	86.96
3.4	90	5	87.72	86.19	71.10	76.09	76.09	80.31	87.47	86.83

12.4.2 Computational cost

As mentioned before, computational costs can be measured by the run times of compression and reconstruction operations. For compression with DCT, since each signal is processed separately, the run times are reported for each time series of the participants, separately. Since we have 17 subjects and 23 trials in which sound stimuli are delivered, the run times should be summed up to calculate the total computational cost. For the CRs reported in Table 12.2, (i.e., 2.2, 2.4, 2.8, 3.4) the corresponding run times for compression were 1.2, 0.9, 0.5 and 0.4 ms. So the run times were reduced as the number of DCTs selected in the compression were lower. However, for each of these compressions, the reconstruction run times were the same: 0.3 ms. For the entire dataset which consisted of 17 participants and 23 trials, the run times of the compressions changed from 469.2 ms to 156.4 ms as the CRs increased. The run times of the reconstruction was fixed for all compressions: 117.3 ms.

During compression with PCA, all of the 23 time series were concatenated and all of the participants were represented in the columns of the matrix, hence, the compression of the entire set of time series for all subjects was done all at once. Similarly, the reconstruction was also done in one matrix operation, once the number of eigenvalues are chosen. Therefore, the runtime for all participants and all trials is measured once for each CR. It took approximately 22 ms to run the "pca" function of MATLAB. Reconstruction is quite faster, and took approximately 5 ms for each CR.

When we compare these runtimes to the ones in DCT compression, we notice that the runtime of the PCA compression and reconstruction is almost 20 times faster than DCT. However, in practice, the application areas of DTC and PCA compression might be quite different. This is because PCA only works in batch mode, once for all subjects and all trials, but, DCT allows online applications in real-time, since each time series can be compressed and reconstructed just by itself, at the receiver and the sender sites.

12.5 Discussion and Conclusion

The variety of the EMG application areas and developments of telemedicine emphasize the importance of the storage and transmission of the EMG data. Novel methods focusing on compression have become a top priority in this regard. Compressing data will ensure minimizing transmission power and saving bandwidth. However, data loss based on compression, latency in transmission, and computational power will stand as drawbacks on the other side of the leverage that should be balanced via optimization techniques.

In this chapter, we focused on the cost and performance issues during compression of EMG signals that are utilized, specifically, for classification applications. While error rates in the signal are of importance in other applications, in our study, we focused on the performance issues based on classification accuracy instead of minimizing losses during reconstruction. The application that we chose was the classification of facial EMG responses to fearful sound stimuli, which can be categorized as a binary classification task.

When performance issues are considered, we can fairly say that compression of facial EMG signals with PCA was favorable. By choosing only 5 eigenvalues, we can reconstruct the EMG signals with a compression rate of 3.4, keeping at least 90% of the variance, while achieving classification accuracy in the range of 86–87% with statistical and frequency based features. If we reduced the compression ratio to 2.1, then we would have to use 8 eigenvalues, but, this would allow us to achieve a classification accuracy above 91% with a 95% reconstruction rate. Such a high reconstruction rate allows for the use of the reconstructed EMG signals in applications other than classification. Hence, compression with PCA at a CR of 2.1 is versatile. Furthermore, the cost of PCA compression in terms of computation time is quite agreeable. Because it takes approximately 20 ms to compress about 400 segments of the EMG time series. Unfortunately, PCA works in batch mode, restricting its application for classification of individual time series on the fly in real time.

On the other hand, compression with DCT proves to be beneficial in real-time applications since it allows for signal compression on the fly in real-time, with an efficient accuracy rate of 82–86% when all features are used. This accuracy rate is the same as the classification accuracy of the original EMG signals, and the computational cost of 0.4–1.2 ms per time segment is quite acceptable to apply compression and reconstruction in real-time. In comparison to compression with PCA, the rates of accuracy in classification are lower in compression with DCT. In addition, the computational costs of compression with DCT are higher than those of PCA. However, regardless of this, the accuracy rates of the classification of EMG signals after compression with DCT are no less than those of the original EMG signals. So, therefore, the benefits of compression outweigh the small operational costs related to computation time.

In short, we investigated whether a biological signal, EMG, is suitable to be compressed for efficient use and transport in a classification application involving effective stimuli. We found that lossy compression and reconstruction of EMG signals is possible without losing much in classification accuracy. The benefits of compression revealed by the compression ratios outweighed the computational costs. Machine learning applications in effective biological signals will become more popular in the future and our study provides clues for efficient use of such signals in both online and offline applications.

Acknowledgment

The authors thank to the Scientific and Technological Research Council of Turkey project for their financial support (Project No: 117E650).

References

[1] Positioning the electrodes on Orbicularis Oculi Muscle. Url = https://web.archive.org/web/20210503154817/http://www.psylab.com:80/html/default_startle.htm.

[2] S. Abhishek and S. Veni. Sparsity enhancing wavelets design for ECG and fetal ECG compression. *Biomedical Signal Processing and Control*, 71: 103082, 2022.

[3] S.A. Alsenan, I.M. Al-Turaiki and A.M. Hafez. Feature extraction methods in quantitative structure–activity relationship modeling: A comparative study. *IEEE Access*, 8: 78737–78752, 2020.

[4] C.P. Antonopoulos and N.S. Voros. Resource efficient data compression algorithms for demanding, wsn based biomedical applications. *Journal of Biomedical Informatics*, 59: 1–14, 2016.

[5] P. Artameeyanant, S. Sultornsanee and K. Chamnongthai. An EMG-based feature extraction method using a normalized weight vertical visibility algorithm for myopathy and neuropathy detection. *SpringerPlus*, 5(1): 1–26, 2016.

[6] M. Awad and R. Khanna. *Efficient Learning Machines: Theories, Concepts, and Applications for Engineers and System Designers*. Springer nature, 2015.

[7] S. Aziz, M.U. Khan, F. Aamir and M.A. Javid. Electromyography (EMG) data-driven load classification using empirical mode decomposition and feature analysis. In *2019 International Conference on Frontiers of Information Technology (FIT)*, pp. 272–2725. IEEE, 2019.

[8] L. Brechet, M.-F. Lucas, C. Doncarli and D. Farina. Compression of biomedical signals with mother wavelet optimization and best-basis

wavelet packet selection. *IEEE Transactions on Biomedical Engineering*, 54(12): 2186–2192, 2007.

[9] L. Breiman. Random forests. *Machine Learning*, 45(1): 5–32, 2001.

[10] S.L. Brown and G.E. Schwartz. Relationships between facial electromyography and subjective experience during affective imagery. *Biological Psychology*, 11(1): 49–62, 1980.

[11] E. Carotti, J.C. De Martin, R. Merletti and D. Farina. Matrix-based linear predictive compression of multi-channel surface emg signals. *IEEE International Conference on Acoustics, Speech and Signal Processing*, 493–496. 2008.

[12] E.S.G. Carotti, J.C. De Martin, D. Farina and R. Merletti. Linear predictive coding of myoelectric signals. In *Proceedings.(ICASSP'05). IEEE International Conference on Acoustics, Speech, and Signal Processing*, 5: 629, 2005.

[13] G.C. Cawley and N.L.C. Talbot. Efficient leave-one-out cross-validation of kernel fisher discriminant classifiers. *Pattern Recognition*, 36(11): 2585–2592, 2003.

[14] S. Chandra', A. Sharma and G.K. Singh. A comparative analysis of performance of several wavelet based ECG data compression methodologies. *IRBM*, 42(4): 227–244, 2021.

[15] W.-S. Chen, L. Hsieh and S.-Y. Yuan. High performance data compression method with pattern matching for biomedical ecg and arterial pulse waveforms. *Computer Methods and Programs in Biomedicine*, 74(1): 11–27, 2004.

[16] G.-Y. Cho, G.-Y. Lee and T.-R. Lee. Efficient real-time lossless EMG data transmission to monitor pre-term delivery in a medical information system. *Applied Sciences*, 7(4): 1–14, 2017.

[17] M.V.C. Costa, J.L.A. de Carvalho, P. de Azevedo Berger, A.F. da Rocha and F.A. de Oliveira Nascimento. Compression of surface electromyographic signals using two-dimensional techniques. *INTECH*, 17–38, 2009.

[18] H. Cui, W. Zhong, M. Zhu, N. Jiang, X. Huang, K. Lan, L. Hu, S. Chen, Z. Yang, H. Yu and P. Li. Facial electromyography mapping in healthy and bell's palsy subjects: A high-density surface EMG study. volume 2020, pp. 3662–3665, 07, 2020.

[19] A.B.M. Sayeed Ud Doulah. Feature extraction scheme based on spectro-temporal analysis of motor unit action potential of EMG signal for

neuromuscular disease classification. Master of Science Thesis, Department of Electrical and Electronic Engineering. Bangladesh University, 2013.

[20] J. Fan and Q. Yao. *Nonlinear Time Series: Nonparametric and Parametric Methods*. Springer Science & Business Media, 2008.

[21] A.J. Fridlund and J.T. Cacioppo. Guidelines for human electromyographic research. *Psychophysiology*, 23(5): 567–589, 1986.

[22] L.J. Hadjileontiadis. *Biosignals and Compression Standards*, pp. 277–292. Springer US, Boston, MA, 2006.

[23] M. Hamedi, S. Salleh, M. Astaraki and A. Noor. Emg-based facial gesture recognition through versatile elliptic basis function neural network. *Biomedical Engineering Online*, 12: 73, 07 2013.

[24] U. Hess, R. Arslan, H. Mauersberger, C. Blaison, M. Dufner, J. Denissen and M. Ziegler. Reliability of surface facial electromyography: Reliability of facial emg. *Psychophysiology*, 54: 12–23, 01 2017.

[25] J.A. Norris, K. Englehart and D. Lovely. Steady-state and dynamic myoelectric signal compression using embeddedzero-tree wavelets. In *Proceedings of the 23rd Annual International Conference of the IEEE Engineering in Medicine Biology Society*, pp. 1879–1882, Istanbul, Turkey, 2001.

[26] N. Jain and V. Renu. Wavelet based vector quantization with treecode vectors for emg signal compression. *Journal of Scientific and Industrial Research*, 67: 117–124, 01 2007.

[27] C.K. Jha and M.H. Kolekar. Tunable q-wavelet based ecg data compression with validation using cardiac arrhythmia patterns. *Biomedical Signal Processing and Control*, 66: 102464, 2021.

[28] S.A. Khayam. The discrete cosine transform (DCT): Theory and application. *Michigan State University*, 114: 1–31, 2003.

[29] S. Khemka, A. Tzovara, S. Gerster, B.B. Quednow and D.R. Bach. Modeling startle eyeblink electromyogram to assess fear learning. *Psychophysiology*, 54(2): 204–214, 2017.

[30] S. Khemka, A. Tzovara, S. Gerster, B.B. Quednow and D.R. Bach. PsPM-SMD: SCR, EMG, ECG, and respiration measurement in response to auditory startle probes. June 2019. Data are stored as .mat files for use with MATLAB (The MathWorks Inc., Natick, USA) in a format readable by the PsPM toolbox (pspm.sourceforge.net).

[31] L.I. Kuncheva. *Combining Pattern Classifiers: Methods and Algorithms*. John Wiley & Sons, 2014.

[32] N. Malešević, D. Markovic, G. Kanitz, M. Controzzi, C. Cipriani and C. Antfolk. Vector autoregressive hierarchical hidden markov models for extracting finger movements using multichannel surface EMG signals. *Complexity*, 2018: 1–12, 02 2018.

[33] W. Marras. Overview of electromyography in ergonomics. *Proceedings of the Human Factors and Ergonomics Society Annual Meeting*, 44: 5–534, 07 2000.

[34] N. Nazmi, M.A.A. Rahman, S.-I. Yamamoto, S.A. Ahmad, H. Zamzuri and S.A. Mazlan. A review of classification techniques of EMG signals during isotonic and isometric contractions. *Sensors*, 16(8): 1304, 2016.

[35] H.R. Neumann and D.L. Westbury. The psychophysiological measurement of empathy. pp. 119–142.

[36] S. Orguc, H.S. Khurana, K.M. Stankovic, H.S. Leel and A.P. Chandrakasan. Emg-based real time facial gesture recognition for stress monitoring. In *2018 40th Annual International Conference of the IEEE Engineering in Medicine and Biology Society (EMBC)*, pp. 2651–2654, 2018.

[37] A.J. Oyobé-Okassa, D.A. Abessolo and P. Elé. Compression of emg signals by superimposing methods: Case of WPT and DCT. *International Journal of Engineering and Technology*, 8: 2319–8613, 2016.

[38] C.K. Park, S.H. Lim, S.H. Lee and B.J. Park. Is the pre-operative lateral spread response on facial electromyography a valid diagnostic tool for hemifacial spasm? *Neurosurgical Review*, February 2021.

[39] N. Pascal, C. Welba and P. Ele. Comparison study of emg signals compression by methods transform using vector quantization, spiht and arithmetic coding. *SpringerPlus*, 5, 12 2016.

[40] S.A. Pavlopoulos, A.C.H. Stasis and E.N. Loukis. A decision tree-based method for the differential diagnosis of aortic stenosis from mitral regurgitation using heart sounds. *BioMedical Engineering OnLine*, 3(1): 21, Jun 2004.

[41] V. Ravi, P.J. Reddy and H.-J. Zimmermann. Pattern classification with principal component analysis and fuzzy rule bases. *European Journal of Operational Research*, 126(3): 526–533, 2000.

[42] G.L. Read. Facial electromyography (EMG). *The International Encyclopedia of Communication Research Methods*, pp. 1–10, 2017.

[43] L. Reidy, D. Chan, C. Nduka and H. Gunes. Facial electromyography-based adaptive virtual reality gaming for cognitive training. In *Proceedings of the 2020 International Conference on Multimodal Interaction*, ICMI '20, pp. 174–183, New York, NY, USA, 2020. Association for Computing Machinery.

[44] K. Rymarczyk, Ł. Żurawski, K. Jankowiak-Siuda and I. Szatkowska. Neural correlates of facial mimicry: Simultaneous measurements of emg and bold responses during perception of dynamic compared to static facial expressions. *Frontiers in Psychology*, 9: 52, 2018.

[45] S.R. Safavian and D. Landgrebe. A survey of decision tree classifier methodology. *IEEE Transactions on Systems, Man, and Cybernetics*, 21(3): 660–674, 1991.

[46] H.S. Khamis, K.W. Cheruiyot and S. Kimani. Application of k-nearest neighbour classification in medical data mining. *International Journal of Information and Communication Technology Research*, 4(4), 2014.

[47] E. Smets, W. De Raedt and C. Van Hoof. Into the wild: The challenges of physiological stress detection in laboratory and ambulatory settings. *IEEE Journal of Biomedical and Health Informatics*, 23(2): 463–473, 2019.

[48] Inc. The MathWorks. *Statistics and Machine Learning Toolbox*. Natick, Massachusetts, United State, 2017.

[49] M. Wauters and M. Vanhoucke. A nearest neighbour extension to project duration forecasting with artificial intelligence. *European Journal of Operational Research*, 259(3): 1097–1111, 2017.

Index

M

N

P

Q

R

S

T

U